IMAGES
of America

ORINDA

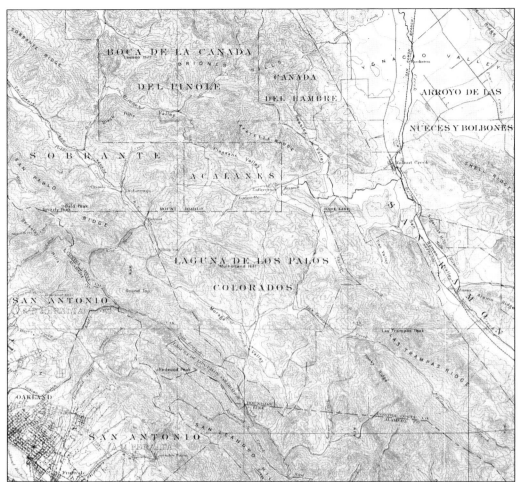

ORINDA, VICINITY OF CALIFORNIA CONCORD QUADRANGLE. By 1897, when this map was produced, Orinda was sufficiently well established to show its three stations along the ill-fated California & Nevada Railroad. The topographic mapping program, which started in 1884, aimed to provide an idea of the country's physical and cultural features. (US Geological Survey.)

ON THE COVER: ORINDA FIRE PROTECTION DISTRICT. In 1933, the members of Orinda's volunteer fire department hung up their helmets and turned over almost $10,000 worth of equipment, including the 1923 firehouse, to the new Orinda Fire Protection District. This delightful scene was captured by Les Sipes, a photojournalist with the *Oakland Tribune*, during the 1940s and 1950s. (Contra Costa County Historical Society.)

IMAGES
of America

ORINDA

Alison Burns

ARCADIA
PUBLISHING

Published by Arcadia Publishing
Charleston, South Carolina

Printed in the United States of America

Library of Congress Control Number: 2022937203

For all general information, please contact Arcadia Publishing:
Telephone 843-853-2070
Fax 843-853-0044
E-mail sales@arcadiapublishing.com
For customer service and orders:
Toll-Free 1-888-313-2665

Visit us on the Internet at www.arcadiapublishing.com

*To the adventurers who shaped this city and the children
who still have the wonder ahead of them as they grow
to treasure this special place called Orinda.*

CONTENTS

Acknowledgments   6

Introduction   7

1.   Natives and Explorers   9

2.   Orinda Begins   19

3.   Selling the Climate   33

4.   Getting Around   51

5.   Schools and Pools   71

6.   Earth, Wind, and Fire   83

7.   Going Out, Staying In   93

Index   126

About the Orinda Historical Society   127

ACKNOWLEDGMENTS

So many have helped with the writing of this book, but I especially thank Mary McCosker (from the Lafayette Historical Society), Susan Sperry and Bonnie Krames (from the Moraga Historical Society), and Janet Stapleton and Mike McCarron (from the Contra Costa County Historical Society) for their kindness and generosity. Thanks also go to my friends at the Orinda Historical Society.

I am also very grateful to the following and wish I had space to do more than simply list their names, but to all of them, I extend my sincere thanks: Bobbie Landers, Amy Worth, James Wright, Susan Leech, John Huseby, Adrian Levy, Ron Kyutoku, Brian Thomas, Alicia Trost, Kathryn Horn, Christina Wong, Toris Jaeger, Pat Adams, Doc Hale, Sora O'Doherty, Sally Hogarty, William White III, Bill Brobeck, Lora Ellen Landregan, Gerald Shmavonian, Claudia Tata, Arlene White, Debbie Jamieson, Kathy Frenklach, Johnny Kirby, Paul Garbarino, Laura and Andrew D'Anna, and Joanna Guidotti. Also, Bob Stoops, Connie de Laveaga Stoops, and Marty de Laveaga Stewart, whose donation of so many unique photographs forms an essential part of the Orinda Historical Society's collections. Thanks also to Mark Harrigan and everyone in the Facebook group "You know you're from Orinda if…," who dipped in and out of the site with their wonderful memories of growing up in Orinda. Meeting them all has been a privilege.

But most of all, I send huge thanks to Reg Barrett, Teresa Long, and Kay Norman, without whose friendship, expertise, and support I would not have gotten this far—and finally, to my husband, Patrick, proofreader and unwavering cheerleader. Thank you all.

All images credited to "CCCHS" are part of the Les Sipes Collection at the Contra Costa County Historical Society. All of the de Laveaga/Orinda Country Club images in this book are part of the collection generously donated to the Orinda Historical Society by the de Laveaga family. Unless otherwise noted, all images appear courtesy of the Orinda Historical Society.

INTRODUCTION

A 1937 real estate pamphlet declares: "If you prefer sunshine and flowers to car lines and crowds, Orinda will claim you—as it will hundreds of others—when they realize that it possesses every rural advantage, yet is still within easy reach of Berkeley and Oakland."

The same promotional materials raved about the "stupendous improvements" that had finally made this land of milk and honey so easily accessible. The Bay Bridge had opened the previous year, and by November 1937, both Contra Costa and Alameda Counties were celebrating the new twin-bore low-level Broadway Tunnel that connected them, which was renamed in 1960 to honor former Berkeley mayor Thomas E. Caldecott.

Within 50 years of California entering the Union as the 31st state in 1850, Orinda's Native American population had been forced out of the area, and the vast Mexican land grants had been forever fragmented. However, without an easy way of accessing the bigger cities, Orinda remained a largely agricultural area.

The idea of a tunnel through the Berkeley Hills had been proposed and rejected in 1860 only to be revived—and rejected again—in 1871. Finally, in 1903, the tunnel became a reality. The narrow, timber-lined Kennedy Tunnel was a vast improvement on the previous means of traveling to and from Orinda but was not for the faint-hearted; until the installation of electric lights in 1914, the only way one could see oncoming traffic in the one-lane tunnel was to travel with a lighted newspaper held aloft.

A 1920s sales brochure advertised Orinda as a modern utopia where "father plays golf as long as he wishes . . . and mother, free from all household cares, follows her own fancies." At that time, Orinda was mostly promoted as a place to build one's country home, although it was also suggested that "some property owners will even commute daily."

In the late 1930s, with the new bridge and tunnel making the suburbs an easy drive from the cities, one brochure boasted that "the pleasures of City life are only twelve minutes away." With the Depression firmly in the rearview mirror, Orinda's realtors began promoting Contra Costa County as a place "where the advantages are great and the taxes small." The trend of travel, said one developer, was toward Orinda, "and where people go, people stay."

And stay they did. Orinda's population has increased more than fourfold since it was at 4,712 in 1960, but most people who grew up in Orinda around this time will allude to a perfect childhood—even the Cold War seems not to have impinged on their memories. Fathers commuted into the city via Greyhound, mothers were homemakers, and if a policeman should discover a teenager driving around town with a beer in her hand, he would gently remove the bottle and escort her home without the need to get parents involved.

Although Orinda schools have always had an excellent reputation, the local education system started slowly. The first school, built near Glorietta in 1861, had just one room and one teacher who covered all grades. It was another 20 years before a second school was built at the other end of town, between Wildcat Canyon and San Pablo Dam Roads. Despite the town's steady growth during the 1920s, school attendance was just high enough to qualify for state aid, and as soon as Wall Street crashed, many of Orinda's residents defaulted on their house purchases and left the area, taking their children with them. It was only when a widow with four children was hired as the school custodian that Orinda's schools were able to qualify.

By 1933, the nadir of the Depression era, Orinda's schools recorded just 26 pupils. With the opening of the tunnel in 1937, that figure almost tripled, and between 1940 and 1953, the size of the student body jumped from 112 to 1,225—an increase of almost 1,000 percent. Suddenly, there was an urgent need for schools in the area. Between 1949 and 1961, Orinda built five elementary schools, two intermediate schools, and one high school.

Incorporation for Orinda had been discussed as far back as 1938, when the population stood at just 767, but did not progress. Two more attempts were made in the 1950s and 1960s, but it was not until 1985, with the population nudging 17,000, that a highly focused committee called Citizens to Preserve Orinda enjoyed a sweeping victory. In July 1985, Orinda officially became a city.

In its early stages, a city is too consumed with the business of evolving to keep tabs on its progress, but as it grows, it begins to look not only at where it is going but where it has been. Orinda's earliest residents, members of the Muwekma Ohlone tribe, left behind artifacts that were unearthed at Lauterwasser Creek and McDonnell's Nursery, just two of the locations where they set up camp. Acalanes, the district to which Orinda's high school belongs, was the name of an Ohlone village known as Ahala-n, which Spanish settlers pronounced as Akalan. Although almost all of the Ohlone peoples were forced out of the area, they still have a voice at the Orinda Country Club golf course, where a large burial site was unearthed in 1924. The 11th hole was named "Graveyard" as a mark of respect, and its fairway was designed with grassy mounds to honor the original burial grounds.

Since 1970, the keeping of Orinda's history has been overseen by the volunteer-based Orinda Historical Society. The Historic Landmark Committee, created in 1987, ensures that places of historic significance continue to be cherished and preserved for future generations. Between 1988 and 2019, the committee successfully adopted 19 landmarks throughout Orinda—some of them simply sites, but all of them marked by informative plaques. There are also a dozen markers around Orinda Village, mostly within walking distance of one another, to guide residents and visitors.

While landmarks and plaques may tell one story about a city, it is the people contained in these pages—in the photographic journey that charts Orinda's unfolding evolution—who have the final word.

One

NATIVES AND EXPLORERS

The Orinda Historical Society's museum has a few precious items on display that originated with the people who lived in the area long before it was called Orinda. These artifacts include an enormously heavy mortar and pestle used by the Ohlone to grind acorns and an 18th-century Spanish cannonball dug up by a 20th-century Orindan who was fixing his backyard drains.

There are plaques around town that tell of the vast, unsustainable acreage bestowed upon grantees by a 19th-century Mexican government eager to displace the Native Americans and populate its newly won lands with its own people. In the late 1700s, when Lt. José Joaquin Moraga established San Francisco, the number of Ohlone people in the Bay Area was around 7,000. By the beginning of the 20th century, they had been virtually wiped out.

Around 300,000 people descended on California during the Gold Rush years of 1848 through 1855, and when the rush was over, the displaced miners filtered down to areas like Lamorinda, where land appeared to be there for the squatting. A handful of Mexican families who owned 13,000 acres could not possibly police their perimeters with so many strangers cutting down their lumber, raising cattle, and ignoring official Mexican credentials.

The bronze plaques around Orinda picture the land grants with straight borders, but these were actually drawn up by vaqueros using ropes to measure boundaries, which made little sense to land-hungry newcomers. Surveyors from the US government redrew the boundaries of these fluid frontiers, and families who had once possessed thousands of acres found themselves having to prove ownership to attorneys who demanded payment in acreage.

THE SAKLAN. An Ohlone tribelet named the Saklan lived in dome-shaped structures fashioned from tule, such as the one pictured at left. This bulrush-type grass was also used for women's skirts, while the men went naked except in very cold weather, when they slapped mud all over themselves. The Saklan spread fishing nets across creek beds and hunted with bows made from plant fiber and animal tendons. When fish and game became scarce, their staple food was acorns ground down by a mortar and pestle, like the one pictured below, which is located at the Orinda History Museum. It was discovered between the Lauterwasser and San Pablo Creeks around 1892, during construction of the California & Nevada Railroad. Skeletons, shells, arrowheads, and other artifacts were also unearthed at Glorietta, Orinda Way, and McDonnell's Nursery on Moraga Way. (Left, author's collection.)

DISINTEGRATION. Once they were forced into San Francisco's Mission Dolores, the native people became severely depressed as they had to endure whippings, disease, and overwork. Louis Choris (1795–1828) whose many paintings documented the Ohlone, said he had never seen one of them laugh. The tribe had numbered around 7,000 in 1770, but under European exploitation, their millennia-old tribal culture rapidly disintegrated. (Oakland Museum of California.)

BURIAL GROUND ON THE GOLF COURSE. In keeping with custom, William Watson, the Scottish-born architect of the Orinda Country Club golf course, gave each of its 18 holes individual names. This photograph looks north toward the 11th hole, located along Orinda Way, which Watson designated "Graveyard" after an Ohlone burial ground was unearthed in 1924. Believing that holes were "found, not built," Watson attempted to honor the burial ground by designing that hole with grassy mounds.

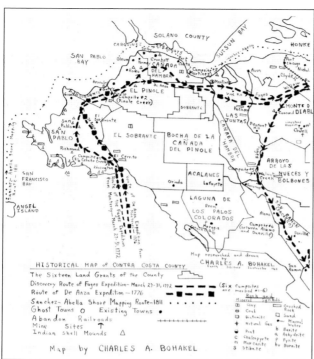

THE ANZA EXPEDITION. In addition to delineating Mexican land grants within Contra Costa Country, the map at left shows the route taken by Juan Bautista de Anza on his legendary expedition to Alta California in 1775–1776. Ultimately, Anza established Spain's overland route to California, but it was his second-in-command, Lt. José Joaquin Moraga, who founded the San Francisco settlement and whose descendants were granted land that later became Orinda. In 1821, Mexico gained independence from Spain. Land grants made under the 1824 Mexican Federal Law covered more than half the total land area of Contra Costa County. The below map gives a sense of the position of the Lamorinda land grants within the county. (Both, Lafayette Historical Society.)

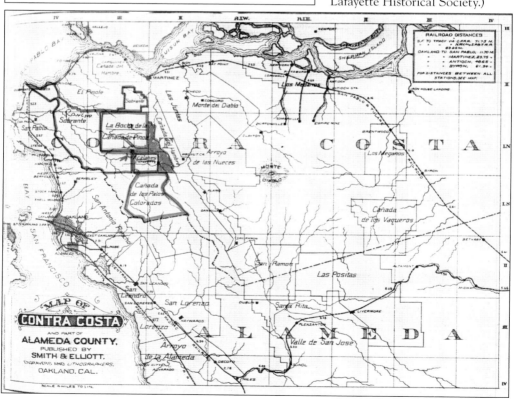

DEL NORTE AND DEL SUR BOUNDARIES. The plaque at right, which is located at Dalewood Park, marks the intersecting location of Rancho Acalanes, Rancho Boca de la Cañada del Pinole, and Rancho El Sobrante. The below plaque, marking the junction of Rancho Acalanes, Rancho El Sobrante, and Rancho Laguna de Los Palos Colorados, is located on Hillcrest Drive. The symbols depicted in bronze represent each rancho's cattle brand. Almost two centuries later, the Tres Ranchos del Sur point is still used as the location for the Mount Diablo baseline—an imaginary line that passes through the intersection of Orinda Way and Santa Maria Way. (Both, author's collection.)

THE JOAQUIN MORAGA ADOBE. In August 1835, Joaquin Moraga, grandson of Lt. José Joaquin Moraga, and his cousin Juan Bernal, whose grandfather had also served in the Anza Expedition, applied to the Mexican government for the three leagues of land that largely make up today's Lamorinda and Canyon. By October of that year, those 13,326 acres were theirs. Today, the adobe that Joaquin built high on a knoll in Orinda in 1841 is celebrated as the oldest building in Contra Costa County. For a while, life was good for the Mexican grantees, but even before California became the 31st state in the Union, squatters were swarming the ranchos and laying claim to land to which they had no right. The Mexican rancheros, who were largely illiterate and unable to comprehend US law, soon found themselves outwitted by opportunist attorneys, and by 1935, Joaquin Moraga's splendid adobe was condemned, dilapidated, and abandoned.

PLUS ÇA CHANGE.
In 1841, Joaquin
Moraga and Juan
Bernal named their
land grant Rancho
Laguna de los Palos
des Colorados
(Ranch of the Lake
of the Redwoods)
in honor of the
pond and trees
that sit on what is
now the campus
of Campolindo
High School.
Despite 100 years
of construction and
clearance, the shape
of the hills in this
1941 photograph
remains unchanged.

DAIRYMAN NUNES. Like so many Portuguese immigrants renting the Moraga Adobe, Joseph Nunes supported his family through dairy work starting in the early 1900s. However, in 1917, when the Moraga Company insisted he plant hay and oats and submit his rent in grain rather than cash, he had to leave Orinda. In this 1930s photograph, Nunes may have been contemplating the life he once enjoyed. (Moraga Historical Society.)

15

THE MORAGA ADOBE REVIVED. In 1941, Katharine White Brown Irvine undertook a total renovation of the adobe despite a 1938 survey warning that while the view was "magnificent," the adobe was in a "deplorable state of repair": the roof needed reshingling, and all the floors, doors, windows, and ceilings needed to be replaced. The surveyor feared that "the labor cost may run to about $1,800," plus $1,200 for materials. He also said that "the historical significance of the Moraga Adobe is not very great since it was not built until the 1840s." However, once the renovation was completed, the adobe became a happy family home, especially when Katharine's grandson, William White III, visited during the years the US Navy was maneuvering PT boats through World War II battles.

THE MORAGA ADOBE

FEATURES

- Two homes –

  the original Adobe built in 1841 and re-
  stored in the 1940's . . . 3,023 square feet of
  living space including four bedrooms, two
  baths, formal dining room, custom designed
  kitchen, library and spacious master suite
  with fireplace.

  separate caretaker's residence . . . three bedrooms, one and one-half baths.

- Heated and filtered swimming pool, with
  adjacent bath house, in a secluded garden
  setting
- Barn with five box stalls and five-ton capacity hayloft
- Irrigated pasture lands
- Walnut grove
- 6,000 gallon capacity spring
- Conveniently located within walking distance of elementary, intermediate and high
  schools.

Offered at $300,000

*Shown by appointment only. For additional in-
formation call the listing broker*

Adele Harlan REAL ESTATE

31 Moraga Way
across from the Orinda Theater
Orinda, California 94563
(415) 254-1544

THE FUTURE ASSURED. After further renovation in the 1960s, the Moraga Adobe went on the market once again, this time with a price tag of $300,000; at this point, it was a registered California historical landmark. But as time went by, it slowly deteriorated. In 2008, the adobe's remaining 20 acres were sold to developers for a subdivision named J&J Ranch (after Joaquin Moraga and Juan Bernal). Although the plan was to turn the adobe into a clubhouse for the homeowners in the subdivision, the newly formed Friends of the Joaquin Moraga Adobe (FJMA) struck a deal with the developers—if FJMA could raise $500,000 by September 2021, the developers would rehabilitate the adobe to its 1848 configuration within a 2.3-acre park. It took 10 years for FJMA to reach the goal, but it was achieved, and the oldest adobe in Contra Costa County will now function as a museum and history center.

THE SULLIVAN RANCH. The Sullivans, a large and energetic family, were part of early Orinda. The first to arrive was Patrick, who was followed in 1879 by his brother, Eugene, who farmed 250 acres around Wildcat Canyon. In 1881, Patrick was found shot to death after accusing his neighbor, Robert Lyle, of stealing his pigs. Although all the evidence pointed to Lyle as the perpetrator, he was freed within weeks of Patrick's death. In 1913, Eugene's son, Jim, married Florence McNeil, the Orinda Park School teacher, and together they ran the Sullivan Dairy from 1932 to 1938. However, they struggled to compete with the larger creameries after the Broadway (later Caldecott) Tunnel opened in 1937.

Two

ORINDA BEGINS

Once the large Mexican land grants disintegrated and new surveys were drawn up, a rush of newcomers arrived, eager to mortgage themselves to thousands of acres in this mostly undeveloped land. Among the first settlers to arrive were William and Alice Camron, whose naming of their less-than-3,000-acre tract eventually applied to Orinda's entire 8,000 acres.

Boundaries were still contentious in the 1870s. Even if a person or family had spent many years farming hundreds of acres, they could still lose most of it once new claims were successfully challenged. In 1874, Augustus Charles, of Charles Hill fame, paid $2.50 per acre for 600 acres only to discover, some years later, that very little of that land was rightfully his.

Water was an issue—the earliest landowners lived close to streams and creeks, but it was obvious that this could not last forever. Water companies started to buy up valuable watersheds, displacing farmers and rarely compensating them adequately for the land they forcibly acquired.

For a while, Orinda had a train service. Today, just one wooden trestle survives—a 21st-century witness to an 1890s railroad that held such promise yet was ultimately so unreliable. Although it was originally planned to go all the way to Nevada (and possibly even Utah), the California & Nevada Railroad never got beyond Orinda.

Many of Orinda's early residents were unable to maintain their flamboyant lifestyles and watched their dreams die thanks to naïveté and bad management, but their names—including Wagner, Bryant, and Lauterwasser—are still spread across the area.

How Orinda Got Its Name. Alice Marsh was the only legitimate child of Dr. John Marsh, a wealthy rancher who was hugely influential in establishing California's statehood. After her father was murdered, Alice inherited his poetry books and was so captivated by Katherine Fowler Philips, a poet known to her friends as "Matchless Orinda," that when she and her husband, William Camron, purchased a Mexican land grant in 1876, they named their new subdivision Orinda Park. However, William speculated badly, and in 1881, he was forced to hold a bankruptcy sale of his Orinda holdings. By 1883, he had squandered his wife's inheritance and deserted Alice and their daughter, Amy. In 1896, Alice divorced William and moved to San Francisco, where she and Amy ran a boardinghouse. Alice never returned to Orinda. In 1927, she was buried in Mountain View Cemetery in Oakland, where Amy joined her in 1963.

MATCHLESS ORINDA. Katherine Philips, the original Matchless Orinda, was an exceptionally well-educated 17th-century Anglo-Welsh poet whose home was the center of the Society of Friendship. In Greek mythology, the word Orinda (a translation of *Ólynthos*) means "wild fig." Philips was also the inspiration for the figure of Orinda in the 1671 Italian tragedy *Il Cromuele*, which is set during the English Civil War.

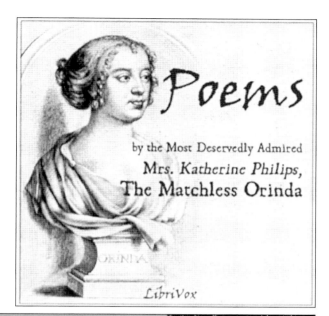

THE CAMRON-STANFORD HOUSE. In 1887, Jose and Miguel de Laveaga paid $50,880 for 1,178 acres in Orinda Park after Alice and William Camron declared bankruptcy. The Camrons' most enduring legacy was the name they gave to their subdivision (Orinda Park) that was destined to become a city, but they also owned the last remaining Victorian mansion on Lake Merritt, which now operates as a museum. (Author's collection.)

GEN. THEODORE WAGNER (1843–1916). German-born Theodore Wagner initially served on the Union side during the American Civil War as an orderly sergeant. In December 1864, having passed his October examination, he was commissioned "in the position of 2nd Lieut. of Infantry, 2nd class, in a US Colored Regiment." In 1873, Wagner, now a qualified attorney, moved to California, and two years later, he was admitted to the state's supreme court. Although he was addressed as "General Wagner" for the rest of his life, this had nothing to do with his military career but rather was the title granted to him in 1878 upon his appointment as the surveyor-general of California. He is pictured in 1900 with one of his grandchildren.

OAK VIEW RANCH, ORINDA PARK. In 1880, Herman Sandow bought 350 acres of William Camron's Orinda Park tract, gifting 241 acres to his daughter Ida after she married Theodore Wagner. The newlyweds named their property Oak View Ranch, and it was located at the intersection of Wildcat Canyon, Bear Creek, and San Pablo Dam Roads. In 1882, they spent around $140,000 building a spectacular, self-sustaining showpiece home with a dairy, an orchard, a blacksmith's forge, a vineyard, a kiln, and a carbide gas plant. They also installed Orinda's first telephone, which had a direct line to Berkeley. Ida raised Theodore's daughters, Lulu and Carrie, as her own, as well as a young relative, Willie Sauer. Before long, their family was completed with baby George, pictured with Ida and Theodore Wagner.

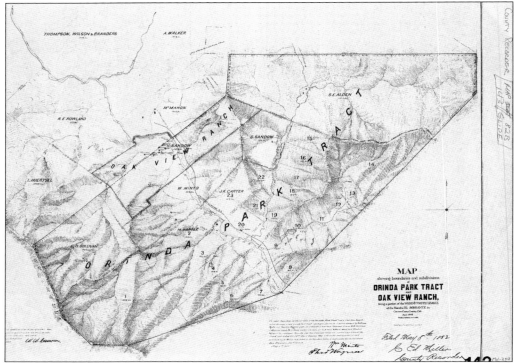

THE MARRIAGE MARKET. After the breakup of the Mexican land grants, speculators started buying large tracts of land, as shown on this 1882 map. Since neighbors lived so far apart, social events were essential for anyone looking to marry. With Orinda's winter storms regularly turning roads into quagmires, it was more fun to dance until dawn than try to struggle home along rutted paths in the dark. One by one, a generation of newcomers married their neighbors: Rudolf Ehlers married Emma Brockhurst (pictured); Theo Wagner married Ida Sandow; the widowed Mrs. Nicholas Brenzel, left with five children after her husband was murdered, married Jacob Ehlers and went on to have another seven babies; Dora Warnecke married George Sandow, and Florence McNeil married James Sullivan. New dynasties were beginning.

FOURTH OF JULY, 1895. The Wagners were famously generous hosts, which may have been Theodore Wagner's downfall, as he was a larger-than-life character with boundless generosity but little business sense. His wife, Ida, must have known this, as an inventory discovered at Oakland's Mormon Temple in the 1990s showed that in an effort to avoid Theo's debts, she listed her own possessions separately from his. The Wagners' "Natal Day" invitation promised the "partaking of luncheon, appropriate poems and patriotic songs," and the local newspaper, the *Contra Costa Gazette*, opined that "the well-known free-hearted hospitality always dispensed at the General's mansion guarantees that guests will be well entertained." The day always finished with a spectacular firework display. However, as the 19th century drew to a close, so did the sybaritic lifestyle. How many attendees at that Wagner celebration guessed that this might be their last?

2

The foundation was commenced on August 15th 1887 and the building is to be finished complete on Thanksgiving day 1887.

Our family consists at this time of the following

Theodore Wagner,
Ida Wagner.
Carrie Wagner.
Louise Wagner.
George J. Wagner.
Willie Sauer.

END OF AN ERA. On July 10, 1887, a kitchen fire burned the Wagner house to the ground, but by August, the family was already laying the cornerstone for their new home, and they were determined to finish it before Thanksgiving. In the above photograph, Ida and Theo Wagner (standing at left) are surrounded by their children and all the Orinda Park School students. In 1895, the impoverished Wagners sold their property to what eventually became the East Bay Water Company. Today, only Wagner School and the Wagner Ranch Nature Area remain. In 1950, a sealed container was discovered in the house's foundations; it contained a bottle of wine, a packet of seeds, and a five-page account (pictured at left) written by Theo that detailed the construction of the new house and the guests at the cornerstone ceremony.

Tug of War. The Fleitz family, pictured above, farmed a 100-acre ranch in Bear Valley during the 1880s and were happy to throw themselves into Fourth of July celebrations. Like so many of those who arrived in the area in the mid-19th century, they were obliged to sell their homes, often at a price they deemed unfair, when water companies started buying up valuable watersheds for the Mokelumne River Project, which ran water from the Pardee Reservoir. The below picture, dated 1916, shows a Marion Model 41 steam shovel stripping creek banks of soil, which was then compacted and hauled by horse-drawn wagons to the new San Pablo Dam. (Below, East Bay Municipal Urban District.)

SLEEPY HOLLOW BEGINNINGS. In 1950, Gertrude, Edith, and Anita Miner created a scrapbook documenting their lives at their ranch, Brookbank (above), in present-day Sleepy Hollow. The book begins with the following comments: "In 1879 father drove mother and three little girls out to the 612-acre ranch that was to be our home for over 40 years. It was built by Frederick Lauterwasser in 1852. They say its tall French windows were shipped around the Horn." The Miner daughters wrote of their English pioneer father, James Miner (left): "Daddie brought with him his lovely sorrel stallion, Don Victor, and continued to raise horses. When the time came to sell a team, he hated to part with them and the $500 (in $20 pieces) that he brought into the house never compensated for having to part with his darlings. The house was remodeled and added onto: running water from a nearby spring was piped to the house and barns."

MAKING HAY. Sisters Gertrude, Edith, and Anita Miner, who grew up in the late 19th century at a ranch called Brookbank in what is now Sleepy Hollow, wrote in their scrapbook about their lives: "The hay in one of the best years yielded just short of 1000 tons. It was hauled to Oakland in six-horse-team wagons. Other farmers were hauling hay too and you'd find yourself behind a long slow-moving wagon which you managed to pass, only to find another one—there were 20–23 of them every summer morning and deep dust. Our linen dusters were really needed!" The Miner girls also remembered "the many guests [pictured below] at the ranch which father and mother enjoyed. Friends came to hunt and fish. There were quail, cottontail and doves in season, trout in the creek. In the winter, at high water, salmon came up San Pablo Creek and we had many a ten or twelve pound fish that the men got with a pitchfork for a spear."

CALLING TO THE ECHO (ABOVE) AND MOTHER MINER (BELOW). According to sisters Gertrude, Edith, and Anita Miner, who resided on Brookbank Ranch in the late 19th and early 20th century in the area that is now Sleepy Hollow, "On a clear day you could see, looking west, to the Berkeley Hills and Mt. Tamalpais, Mt. St. Helena, and to the north, Mt Diablo. And on a very clear day, the line of snow-capped Sierra Nevada mountains in the far distance. At the corner of the ranch is the surveyor's concrete marker, marking the corner of the Sobrante with the Acalanes grant and Briones." Remembering Anna, their mother, the sisters wrote: "For mother, always a birthday party on May 18th. Those early years were, for my parents, filled with work and the hardships that go with pioneer life. To the child that I was it meant nothing, only the added interest of watching the building of the new barn and the rearranging of the house to make it more livable for mother."

CHANGING TIMES. It is extraordinary to think of the timeline the three Miner girls straddled. Born in the 1870s, when theirs was the only house within 612 acres, Gertrude, Edith, and Anita Miner spent their days in flowing white dresses, traveled dusty unpaved roads on horseback, and as Californian women, had to wait until 1911 to enjoy equal voting rights. By 1955, the year of Edith's passing, the United States' first nuclear-powered submarine was putting out to sea, Elvis was on television, and Jonas Salk had just launched his polio vaccine. In her 1950s scrapbook, Anita Miner marveled that when she and her sisters were growing up, "there was no electricity or telephone, and you cooked almost all your food on a wood stove, on which you heated your sadirons to iron your starched and embroidered undies." In 1909, when James Miner died at the age of 72, his widow, Anna, signed the ranch back to Ann Miner, her sister-in-law. James and Anna's family had enjoyed Brookbank Ranch, but it was getting increasingly difficult to make a living from raising horses. In the 1920s, the property became Sleepy Hollow Syndicate, with just a few houses appearing, but by the 1950s, large-scale housing development was well underway. Today, Sleepy Hollow is considered one of Orinda's most desirable areas, and although the old Brookbank Ranch is long gone, the past is still invoked in names like Miner Road, Lauterwasser Creek, and Brookbank Road.

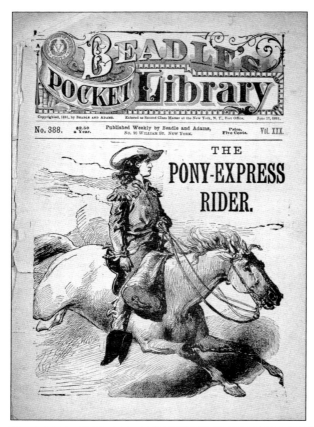

THE PONY EXPRESS. The plaque at the intersection of Camino Pablo and Brookwood Road is dedicated to the Pony Express boys who galloped through Orinda between April 1860 and October 1861 on the final leg of their journey from St. Joseph, Missouri, to San Francisco. Although Orinda was never officially part of the route, on the 20 occasions that severe weather or attacks by Native Americans led to missed connections at the Sacramento ferry, the riders would race overland to Oakland. Legend has it that they had to be "young, skinny, wiry fellows not over eighteen" and take an oath not to "fight, use profanity, or drink intoxicating liquor." One was the legendary Buffalo Bill; perhaps he hastened through Orinda on his way to catch the Oakland ferry? (Below, author's collection.)

Three

SELLING THE CLIMATE

In 1887, Miguel and José de Laveaga, two brothers from San Francisco, paid William and Alice Camron $50,880 for 1,178.04 acres within Orinda Park. A year later, Miguel and his wife, Marie, moved to Orinda and built their country home, Bien Venida.

Miguel instinctively knew that this garden of Eden would eventually become something special and told his son, Edward Ignacio de Laveaga (known as "E.I."), that "the land is becoming too valuable to hold—someday we're going to have to give it away and sell the climate."

In 1921, E.I. embarked on the first of his subdivisions. Aware that he would need to present potential buyers with a reliable water supply, he began by constructing Lake Orinda.

A year earlier, the old wooden bridge that crossed the San Pablo Creek had been replaced, and the muddy track, now called Orinda Way, instantly became a viable location for E.I.'s village. His plans included a firehouse, a riding school, a garage, a hotel, a chapel, a nursery, a theater, and a few stores, all of which he surely would have achieved if the events of 1929 had not sent the real estate market into a free fall. But by then, he had transformed a 19th-century farming community into a desirable place to live.

Today, a modern freeway and transit system run through E.I.'s dream town, and his influence lives on in Orinda's Spanish-themed architecture, its 70 streets with Spanish names, and a golf course that remains one of the most challenging of the classic-era designs in Northern California.

BIEN VENIDA. In 1888, Miguel and Marie de Laveaga built a house off Miner Road and named it Bien Venida—Spanish for "welcome." In 1915, according to family lore, just hours before E.I. de Laveaga (son of Miguel and Marie) planned to ride into San Francisco to purchase fire insurance, the house burned down. Within the year, it had been rebuilt to the original plans. Miguel is pictured below in 1909 with E.I.'s wife, Delight. At that time, cars were still in their infancy and inclined to be far more temperamental than horses, and Delight often complained that despite starting a journey by car, it was not unusual to find herself, halfway through the trip, stranded in a broken-down automobile and waiting to be rescued—by a horse and buggy.

CHURCH OF SANTA MARIA CHAPEL. In 1892, Marie Le Breton, wife of Miguel de Laveaga, built a small chapel a short stroll from their house, Bien Venida. The dedication service took place in October 1896, but Marie, who had recently given birth to a stillborn child, was too sick to attend. Sadly, she joined her baby just days later. By 1954, the chapel was proving too small for the congregation, and a larger church was built on Santa Maria Way. The little chapel was demolished, but its steps are still visible at the beginning of Miner Road, and its bell, designated an Orinda historic landmark in 1995, is located just outside the new church. (Below, author's collection.)

E.I. DE LAVEAGA (1884–1957). The story of Orinda Village is really the story of E.I. de Laveaga, whose vision, planning, and determination truly shaped Orinda. In 1921, he embarked on a breathtaking plan to create a new community where city-dwellers would build weekend and country homes and eventually settle full-time. But first, he had to create the infrastructure, including roads and water. After that came a general store, a garage, riding stables, a swimming lake, and a country club, which he could present to potential purchasers looking to buy an empty lot. E.I. was riding high when the Depression hit, and he watched his fortune melt away as his buyers vanished. His bankers, however, knew him to be an honorable man and trusted that he would one day settle his debts—which he did, entirely, in the 1940s.

HUNTING, SHOOTING, FISHING. E.I. de Laveaga was born a city boy but had spent summers in Orinda since he was four, and he moved there permanently in 1913. The area abounded with coyotes, deer, and wildcats, and E.I. proved to be an excellent marksman. So good was he that when the Oakland Museum was adding to its trophy collection, it recruited him as a huntsman. He was also an accomplished angler, and before the San Pablo Creek was dammed in 1919, he regularly brought home steelhead trout and salmon. These images show hunters near the Sullivan Ranch in the early 1900s, when El Toyonal still offered plenty of challenges.

HAIL THE HORSELESS CARRIAGE. In 1900, when the photograph at left was taken, there was around 1 vehicle for every 100,000 people in the United States. Orinda was still quite isolated, since Oakland was only accessible via a vertiginously steep route across the Oakland-Berkeley hills. The below picture, taken after E.I. de Laveaga's family had traded horsepower for the combustion engine, shows two of E.I.'s children in their father's well-upholstered REO runabout. In the 15 years between these two images, the ratio of cars to people increased more than 200 times, and E.I. had embarked on the subdivisions that would lead to San Franciscans flocking to Orinda.

IT TAKES A VILLAGE. Grading roads in 1924 was a labor-intensive business for both man and beast. It was even harder in 1918, after the construction of the San Pablo Dam. During heavy rains, the creek flooded so badly that access was impossible until Contra Costa County built a new bridge in 1920. The square building on the left was the town's first commercial structure. Built by Ruth and Frank Enos in 1921 and named the White Swan, it housed a small store, refreshment stand, and gas station. E.I. de Laveaga later bought it for use as a bunkhouse for his workers, renaming it Casa Verana. The structure just across the bridge on the right is Orinda's original firehouse, built in 1923. The firehouse, Casa Verana, and the 1920 bridge are all still in existence today.

THE ORINDA SIGN. In the late 1920s, motorists keen to investigate Orinda's new subdivisions knew they had come to the right place when they spied the huge, newly constructed Orinda sign perched over Orinda Way. While the original sign was made of wood, its replacement was fashioned out of green metal some 20 years later. It later graced the Woolsey Real Estate office at the Camino Sobrante and Orinda Way triangle but disappeared when the office was demolished in 1969. Fifteen years later, the sign marked the headquarters for Orinda's incorporation campaign and now sits amongst the shrubs outside the Orinda Community Center. (Below, author's collection.)

ORINDA VILLAGE, 1926. By 1926, E.I. de Laveaga's village was taking shape. The previous year, de Laveaga had opened the Orinda Store (below), which later became Phair's. Anyone wishing to avoid the perilous Kennedy Tunnel into Orinda could park at Grizzly Peak Stables in the Berkeley Hills and continue on horseback to the Orinda Riding Academy. Leaving their mounts in the excellent care of Miss Graham (otherwise known as Mrs. Philip R. Donaldson), prospective purchasers then rented a room at the Orinda Country Club for $40 for the month. Horses could be rented from the riding academy for $1 for the first hour and 50¢ per hour thereafter. (Above, Kirby family.)

LAKE ORINDA. Unlike Orinda's early pioneers, who settled near creeks and streams, E.I. de Laveaga knew it was essential to provide a reliable water supply to prospective homeowners. So, he built Lake Orinda high in the hills now known as El Toyonal (shown in the upper left of this 1924 photograph). The lake subsequently became Orinda Park Pool. Note the construction of what would eventually become Phair's Department Store.

BOATING. Everyone who bought a lot in the Lake Orinda subdivision was guaranteed water—plus one share of the water company. By October 1922, one third of the lots had been sold, and by 1923, every drop of water that could be mustered through pumps, valves, and springs was making its way 1,500 feet uphill to the new subdivision of Orinda Park Terrace.

HACIENDAS DEL ORINDA, 1924. In the office between Mira Loma and Camino Sobrante, E.I. de Laveaga's sales agents were told to "forget real estate—the general idea must be sold to the prospect first." Their handbook also told them to "remember that you are dealing with people to whom courtesy and good manners are most important."

MITCHELL & AUSTIN. Members of the Sorensen family are pictured taking a break outside the Mitchell & Austin Real Estate office. Robert Mitchell had worked as sales manager for the Lake Orinda subdivision since 1921, but in 1925, he and Harold Austin opened new premises at Orinda Way and Camino Sobrante to represent Hacienda Homes, Inc.

CROSSROADS REAL ESTATE. Once the Broadway (now Caldecott) Tunnel opened in 1937, property sales flourished, and the distinctive, red-roofed Wallace office at the Orinda Crossroads, next door to Willoughby's, pictured here in 1941, remained a beloved landmark until it was bulldozed in 1967 to make way for Bay Area Rapid Transit construction. Note the Casa Orinda restaurant, which relocated to its present site in 1942.

MORAGA ESTATES, 1936. For 20 years, Fred T. Wood, based at the Orinda Crossroads, was a hugely successful developer throughout Orinda. But these were less enlightened times, and the terms of the sale in his first subdivision, Moraga Estates, restricted (until 1961) "persons of African or Mongolian descent."

MASSIE & UNDERWOOD, 1955. Before a multilane freeway and rapid transit system altered the orientation of Orinda, the Orinda Crossroads formed a major business area with cocktail bars, restaurants, and well-positioned real estate offices, such as Massie & Underwood. Casa Orinda has remained an immutable fixture, but little else remains of the original businesses.

ONE SHARE. After purchasing a lot in the Hacienda Homes subdivision, each buyer got a certificate stating that Hacienda Homes, Inc., held one share of stock on their behalf in the water company. This insured that control of the water company remained with Hacienda Homes until the property owners functioned as an administrative body.

COUNTRY CLUB REALTY, 1960. In 1960, when this photograph was taken, Country Club Realty could easily afford this triple-fronted office in Orinda Village, since the town was rapidly becoming a desirable place for the upwardly mobile to raise their families. In 1960, according to the Bay Area Census, Orinda's population numbered 4,712, of which 99.5 percent were white. That year, almost 93 percent of Orinda's residents had been born in the United States, with the town's 2,388 women outnumbering men by 64. With a median income of over $10,000, it was easy to see why 96 percent of homes were owner-occupied. Interestingly, in the over-25 age group, only 28 percent had completed four years of high school. Perhaps the most surprising statistic is that while 53 percent of Orindans belonged in the 20–64 age group, only 4.7 percent of the population was over 65.

CONSTRUCTING LAKE CASCADE. In 1923, when E.I. de Laveaga began creating the 580-acre Haciendas del Orinda, his first task was to build a 400-foot dam across an 18.5-acre valley to collect water from springs, wells, and natural drainage. Thus, Lake Cascade was born. It initially held 62,740,000 gallons, with an eventual capacity of 97,740,000 gallons. The sales brochure stated that "in addition to forming one of the principal scenic features of Haciendas del Orinda, Lake Cascade will supply excellent filtered fresh water to all of the homesites and furnish an abundant supply of water to the golf course, making possible the operation of the course at minimum expense."

LOOKING WEST (ABOVE) AND SOUTH (BELOW). Perhaps the current owner of Lot 115 (above) would be interested in the view around Lake Cascade as it was in 1924. The lake was surrounded by a great variety of stunning trees, with an additional 3,000 pines and redwoods planted around the new subdivision. E.I. de Laveaga also maintained a greenhouse near his home where thousands of trees were developed for future planting. Tourists came from San Francisco, the Peninsula, and all over the East Bay area to picnic by Lake Cascade. Boating was permitted only until the lake was needed as a water supply.

HACIENDAS DEL ORINDA FROM THE AIR. This aerial image was taken when the Orinda Country Club golf course was already in full swing but before the tennis courts appeared, and when a stroll around Lake Cascade promised a gentle perambulation. Anyone familiar with Dos Posos today will be amazed to see how solitary it once was.

ORINDA FACTS

(1) Self-contained community with its business center, school, churches, fire department, library, etc.

(2) Suburb of Oakland and Berkeley.

(3) Same Utilities.

(4) Same Rates.

(5) Same Water.

(6) Same Phone Service (a Berkeley exchange).

(7) Low Taxes.

(8) Lots, one-fifth of an acre to 5 acres.

(9) 162 Homes.

(10) 25 Homes built in last two years.

(11) A number of F.H.A. Loans have been approved in this subdivision.

(12) Miles of Roads . . . 20
Miles of Trails . . . 17
Good Schools.
Close to University of California.

(13) Bus Service to be inaugurated to Orinda upon the opening of the Broadway Low-level Tunnel, scheduled for the first part of 1937.

(14) Orinda is one of the fastest growing sections of Contra Costa County. In 1922 the assesed valuation was $25,000.00. In 1936—$775,000.00.

HACIENDA HOMES • Incorporated

Telephone THornwall 1802   **ORINDA, CALIFORNIA**

ORINDA FACTS. In 1937, when E.I. de Laveaga was promoting his town as one of the fastest growing within Contra Costa County, he quoted two figures: Orinda's assessed valuation in 1922, when it was $25,000, and in 1936, when that valuation had risen to $775,000. He would feel validated to learn that by 2021, it had increased to almost $7.8 billion.

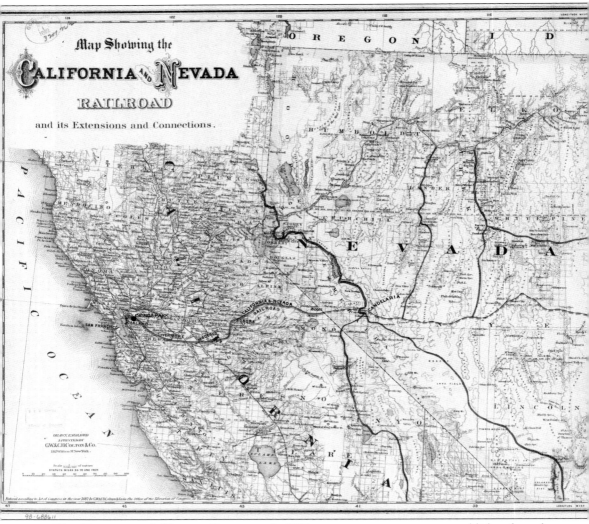

THE CALIFORNIA & NEVADA RAILROAD. In 1881, a company was set up to establish a three-foot, narrow-gauge steam railroad that promised investors a route beginning in Emeryville (known as Emery's at the time for pioneer developer Joseph Emery), which would then continue east across the Sierra Nevada to the gold and silver mining town of Bodie. From there, the line would travel east to connect with the Denver & Rio Grande Railroad in Utah. Although sparsely populated towns such as Orinda Park would not have appeared on this map when it was drawn in 1882, its approximate position is marked near the start of the long horizontal route planned for the California & Nevada. Orinda had three stations: Orinda Park, de Laveaga, and Bryant—named after San Francisco mayor Andrew J. Bryant, who built his summer home in the area near the Orinda Theatre. A brass plaque showing the location of the Bryant Terminus is located just yards from the theater, close to Highway 24, where 170,500 vehicles roar by on a daily basis.

Four

GETTING AROUND

In an article printed in the *Orinda News* on January 13, 1947, George Brockhurst (1871–1963) tells of the time in the early 1900s when his six-horse team got into "a dickens of a snarl" in the Kennedy Tunnel, and he was unable to untangle the lead-span until he chanced upon a fellow walking toward him in the dark with a candle.

In 1937, the new twin-bore Broadway (now Caldecott) Tunnel opened, and Orinda's position as the gateway to Contra Costa County was assured. People had already grown excited at the prospect of settling in the suburbs, and house construction resumed in Orinda as realtors' offices prepared for the population explosion. This modern tunnel was far safer than the previous one—and even more so after a horrendous fire in 1982 led to banning tankers from transporting hazardous materials outside the hours of 3:00 a.m. to 5:00 a.m.

State Highway 24 improved exponentially in the mid-1950s, and many commuters took their cars into the city. Others preferred to let the Greyhound do the driving. Even though a locomotive had occasionally puffed its way into Orinda in the 1890s, it had really only been used by Sunday picnickers heading for the dance floor at the end of the line. However, the train that pulled into the Bay Area Rapid Transit station for the first time on May 21, 1973, was an entirely different beast, and although the bus riders staged a valiant fight to prevent losing Greyhound service, it was clear that a new era had arrived.

ORINDA'S STATIONS. The above image shows fireman George Pope (left) and engineer H. Aleck Robertson (right) in 1885 proudly standing by locomotive No. 3, a coal-burner built in 1884. The below picture, looking north, shows the California & Nevada Railroad terminus at Bryant's Corner around 1895. Note the dance platform and picnic area. Note, too, the old pine grove. This area was known as the Orinda Crossroads before Highway 24 came through. By 1901, the narrow-gauge line to Orinda had been abandoned, but it is interesting that when the Bay Area Rapid Transit system (BART) reached Orinda in 1973, its station was located just a quarter mile from the old California & Nevada Railroad terminus.

A RHEUMATIC LOCOMOTIVE. According to an April 1893 article in the *Hayseed Siftings*, "we have found that a train is very convenient even it if it is not always on time . . . and that a railroad not on time is better than no train at all." Crowds gathered for a "rheumatic locomotive" that rarely adhered to the timetable, but it was much beloved by day-trippers and picnickers and proved useful to farmers transporting hay and harvests to the Emeryville docks. The carriage roofs leaked so badly that passengers sitting inside often spent the entire journey beneath their umbrellas, and roughly once every week, the wood-burning engine set hay fields on fire. (Above, Wayne York Collection.)

DE LAVEAGA DEPOT. Originally located along Miner Road, the California & Nevada Railroad's de Laveaga station was moved in the 1920s, long after the railroad failed, to make way for the new golf course. It was used as a toolshed on the de Laveaga property until 1999, when E.I. de Laveaga's great-grandson Andrew Stewart and his Troop 303 buddies embarked on a 100-hour Eagle Scout project to gently relocate the station close to where the old terminus had once stood at Bryant's Corner. Now, the thoroughly renovated building sits next to a portion of the narrow-gauge track and forms part of the Orinda Historical Society's walking tour.

THE DREAM DIES. In 1882, Chinese laborers bored a tunnel through Charles Hill, and in 1891, the Botillo family donated land for a depot in Walnut Creek. Track was graded along Moraga Way as far as Glorietta in 1893, but the entire enterprise was so beset by disaster that the 23 miles of track from Emeryville to Orinda was where the California & Nevada Railroad dream finally fizzled out. The crew got so used to the train derailing at the notoriously swampy "Frenchman's curve" that they kept blocks and a 20-foot long lever on hand so the wheels could be jacked back into place—with the assistance of half of the passengers. In 1892, Elmer W. Barnes (pictured) was named chief conductor, but by 1895, he was ready to retire.

MAGNIFICENT TRESTLES. In the 1890s photograph above, the de Laveaga family pose with friends on a special excursion train perched on the highest trestle of the narrow-gauge track. The de Laveaga station, located at the far end of the second trestle over Miner Road, is barely visible. After crossing what is now the 18th fairway of the Orinda Country Club golf course, the train would continue past Orinda Village toward Bryant's Corner and deposit picnickers at the dance platform where live music was often played. Now, all that remains of the magnificent structure that once dominated 19th-century Orinda is one timber beam behind the Orinda Community Park. There are also some fascinating information boards—worth seeking out— along the park's footpaths. (Left, author's collection.)

FISH RANCH. In Victorian times, getting from Orinda to Oakland was an onerous journey—even a horse-drawn buggy took two hours, and it was easily double that for farmers driving a team to market with a heavy load. A popular resting place was the Oakland Trout Company's 175-acre fish and frog farm, built in 1872 near the present-day Gateway Boulevard exit off Highway 24. Managed by George Winslow and his wife, Martha, this welcome stop came to be called Fish Ranch and was famous for a hatchery that provided travelers with trout and salmon dinners. After George Winslow passed away in 1879, his son-in-law George Olive took over, adding a tavern and guest cottages where travelers could wind down. The tavern closed in 1915.

THE OFF-RAMP, C. 1890.
Looking remarkably composed
after the bone-shaking journey
over Fish Ranch Road, these
picnickers are about to cross
the San Pablo Creek into
Orinda. The road upon which
they had traveled, located
just behind them, is now an
off-ramp from Highway 24.
Ahead of the party, straw
has been laid by the creek to
allow for better traction.

THE ORINDA CROSSROADS, 1936. As Orinda grew in popularity, it was natural that the crossroads at the intersection of Camino Pablo and Tunnel Road should become a magnet for new businesses. Gus Reuter established the first Standard Oil gas station in the late 1920s, followed by the Crossroads Restaurant, which remained a favorite until it succumbed to the highway-widening project in the 1950s.

Towards Oakland and Tunnel

Now Highway 24

Camino Pablo

Moraga Way

Pine Grove

Northwood Dr.

Orinda Crossroads

Davis Road

Towards Walnut Creek

THE ORINDA CROSSROADS. Although the above photograph is undated, it was taken before the Great Depression was officially over, before the United States entered World War II, and before the Orinda Theatre appeared on that empty lot on the right. The below photograph is dated March 1950, almost 5 years after the end of World War II and 13 years after the Caldecott Tunnel opened up the suburbs. In the 1920s, the general store on Orinda Way had been all the town needed, but when its beauty shop was destroyed by fire in 1939, the owner knew it was time to relocate to Orinda's hottest commercial area: the Orinda Crossroads.

EXPANDING THE HIGHWAY. After it opened in 1937, the Caldecott Tunnel quickly proved to be a victim of its own success: traffic increased in far greater numbers than anticipated, and by 1943, it had become necessary to widen part of Highway 24 between Orinda and Lafayette to four lanes. Even this was not enough, and changes were soon in the pipeline. Within 10 years, all the businesses that had once clustered around the Orinda Crossroads had disappeared, and major work, pictured here in 1955, was undertaken to construct an underpass and a six-lane divided freeway with cloverleaf interchanges. (Both, CCCHS.)

SIGN OF THE TIMES. It may be hard to imagine a time when there was no traffic signal at the intersection of Camino Pablo and Brookwood Road, but it was only on January 12, 1961, that one was first installed. There to help smash the celebratory champagne bottle were Clarence E. Betz (right), secretary of the Orinda Chamber of Commerce, and "Keystone Cop" Bob Hecocks. After the opening of the Broadway (now Caldecott) Tunnel in 1937, Lamorinda residents found a daily commute to San Francisco much more feasible, and by the early 1950s, traffic at Orinda's crossroads numbered more than 31,000 vehicles per day—or around 3,000 cars per hour during peak periods. (Both, CCCHS.)

THE CROSSROADS MARKER. Although the area once known as the Orinda Crossroads is gone, a historical marker placed close to where the BART walkway intersects the station's east parking lot indicates where this vibrant hot spot was once located. Other markers within walking distance include those for the Pony Express, Bryant station, de Laveaga station, Santa Maria chapel, Orinda Library, and the 1923 Orinda Improvement Association. (Author's collection.)

TURNING TURTLE. On this sunny day in June 1960, just west of the Orinda Crossroads, it was difficult to tell whether this upturned vehicle on Highway 24 was going to or coming from Orinda. The badge on the grille identifies it as the long-forgotten Nash Rambler, which some might remember as the little car in the Playmates' 1958 song "Beep Beep." (CCCHS.)

THE KENNEDY TUNNEL. In the 1890s, the Miner sisters—Gertrude, Edith, and Anita—poked fun at their neighbor, Mr. Buckley, whose "constant talk about a tunnel through the hill" led people to regard him as "mildly demented." He was vindicated in 1903, when the hand-dug 1,040-foot-long Kennedy Tunnel connected Oakland with Orinda despite a miscalculation that resulted in a four-foot jog midway through the tunnel. The single-bore Kennedy was narrow, dark, and dangerous. By the time work began in 1934 on the Broadway (now Caldecott) Tunnel, located 300 feet below its predecessor, almost 30,000 vehicles and pedestrians were valiantly struggling through the Kennedy Tunnel each week. The photograph at right shows the old tunnel above the new tunnel's west portal.

TUNNEL OPENING. The new low-level Broadway Tunnel, with two bores measuring 3,203 and 3,135 feet in length, opened on December 5, 1937, with ceremonies featuring planes, fireworks, pigeons, and an exploding wall. Once the speeches were over, the huge crowd surged through the Art Deco portals and raced to the other side. Peggy Kirby was among the lucky few to receive a postcard that reads: "This certifies that the holder was one of the first one thousand persons who passed through the Broadway Low Level Tunnel December 5, 1937, at the opening for public travel." (Below, Kirby family.)

Third and Fourth Bores. In 1960, the Broadway Tunnel was renamed the Caldecott Tunnel in honor of former Berkeley mayor Thomas E. Caldecott. By this time, traffic had grown to over 50,000 vehicles per day, and construction had begun for a third bore. When it opened in 1964, it contained plastic "pop up" lane delineators that reversed the traffic flow according to the demands of commuter traffic. The below photograph shows the relative tranquility of the Caldecott Tunnel in 1966 as it passed from Orinda to Oakland and beyond. The fourth bore, which opened in 2013, contains traffic-monitoring systems, excellent ventilation, electronic message boards, and emergency exits into the third bore. (Both, Caltrans.)

A Perfect Storm. On April 7, 1982, shortly after midnight, a woman driving through the Caldecott Tunnel's third bore struck a curb and stopped to examine the damage. Close behind her, a gasoline tanker careened into her car. Next, a speeding bus hit both vehicles, the driver was thrown clear, and the bus continued through the tunnel and crashed. As fires flared around the tanker, 8,800 gallons of burning gasoline poured through drains and into Lake Temescal. As a 100-mile-per-hour fireball roared through the tunnel, the inferno soared beyond 2,000 degrees Fahrenheit, but the ventilation system only kicked in after carbon dioxide levels exceeded the trigger level. Sadly, it took seven unnecessary deaths to expose the tunnel's shortcomings. (Both, Caltrans.)

Bay Area Rapid Transit (BART) Begins. Francis "Borax" Smith proposed the idea of an underwater electric rail tube in the early 1900s, although San Francisco's Emperor Norton might perhaps argue that he had dreamed of a trans-bay tunnel as early as 1872. It took almost another century for Norton's dream to come true. Construction of the Bay Area Rapid Transit system officially began in Concord on June 19, 1964, with Pres. Lyndon Johnson presiding over the ground-breaking ceremonies for the 4.4-mile Diablo Test Track to Walnut Creek. By 1966, work was well under way to construct a station platform in Orinda between the east- and westbound lanes of the widened and relocated freeway, and by the time Orinda's station opened on May 21, 1973, BART had 24 stations along 56 miles of track. (Both, BART archives.)

No Straphangers. In the late 1960s and early 1970s, Orindans had a unique opportunity to examine the evolving freeway and nearby structures as BART slowly took shape. Over time, a parking lot for 200 cars was added. In 1965, BART issued a series of architectural drawings detailing the line's space-age carriages; the luxurious interiors had seatback maps but—curiously—no straps or bars for standing passengers. When questioned about this, management explained that this was because they hoped to seat everybody, and in any case, the schedule would be staggered so well that a new train would arrive within minutes of the last one. (Both, BART archives.)

BART VS. GREYHOUND. In 1956, Greyhound Bus Lines came up with the slogan "leave the driving to us"—and for 37 years, that is what Orinda's commuters did. However, by 1978, the competition from BART was gaining ground. That year, when Greyhound's management decided to terminate the company's service from Antioch to San Francisco, they found that they had a fight on their hands. Complaints about BART ranged from "stopping too many times" to "still running six minutes apart." The red, blue, and silver buses had been such an icon of American life since 1914 that simply seeing them signal a right turn at the Orinda freeway exit was something many people found hard to relinquish. (Above, Caltrans.)

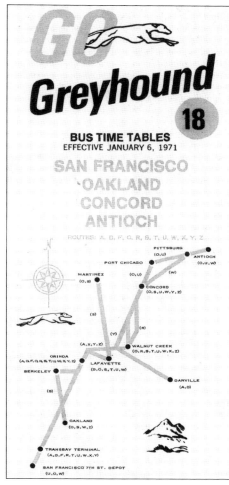

GO Greyhound **18**

BUS TIME TABLES
EFFECTIVE JANUARY 6, 1971

SAN FRANCISCO
OAKLAND
CONCORD
ANTIOCH

ROUTES: A, D, F, G, R, S, T, U, W, X, Y, Z

Free Commuter
PARKING
Courtesy
SHELL OIL COMPANY

THIS CONTRACT LIMITS OUR LIABILITY-PLEASE READ IT

GOODBYE GREYHOUND. Despite paring down the Contra Costa fleet from a pre-BART figure of 120 to 21, Greyhound continued to operate at a loss. By the mid-1970s, Greyhound's weekday costs totaled around $1.90 per mile, of which 64¢ was pure loss. In 1978, when Greyhound decided to terminate its service, riders argued there was still no adequate alternative for the 600 commuters from Orinda, Lafayette, Walnut Creek, and Concord. When the end finally came, many dyed-in-the-wool Greyhound devotees, forced to abandon the 1952 bus shelter donated by the Orinda Chamber of Commerce, chose to carpool into the city rather than take BART. (Both, Caltrans.)

Five

SCHOOLS AND POOLS

In some way, little seems to have changed since Orinda's first school opened in 1861—much like today, there is never quite enough funding, books, or teachers. On the plus side, today's students no longer have to huddle around an old potbellied stove in the winter months or give up school altogether during weeks of heavy storms, when the track to the school became an impassable quagmire. They can also be sure they will have indoor plumbing. On the other hand, it was perhaps more fun for children to do the school run in a horse and buggy than in an SUV.

Orinda's excellent schools have always been a strong draw for parents as they choose where to raise a family. After the opening of the Broadway (now Caldecott) Tunnel in 1937, the rush to the suburbs began. Today, Orinda's enrollment numbers for students from transitional kindergarten through eighth grade are at almost 2,500—a far cry from the 36 students who comprised the head count for Orinda's two schools in 1904.

In the late 1800s and early 1900s, area children spent summer vacations working on the family ranch or farm. These days, they are more likely to be found in a pool. According to CountyOffice.org, Orinda ranks 35th out of 1,798 Californian cities in swimming pools per capita. The Orinda Moraga Pool Association (OMPA) was established in 1956 to run a summer recreational swim league, which concludes with an annual championship meet—and explains why vehicles drive around in August with shark fins and gators wobbling precariously on their roofs.

MORAGA SCHOOL, 1861. Orinda's first school was built in 1861 on Moraga property across from where the Old Yellow House still stands. In 1892, Jennie Bickerstaff (pictured above in 1890) began teaching students in first through ninth grades, and she remained at the school for seven years. Since it was unbecoming for Victorian ladies to straddle a horse, she rode Topsy sidesaddle three miles each way from her home, where Diablo Foods now stands. The saplings she planted remain but are now towering redwoods. When the ground was too muddy for cross-country riding, Bickerstaff took the dirt road (the precursor of Mount Diablo Boulevard), which added another four miles to her daily ride. Once a month, she traveled to Fish Ranch to request her $60 salary and then to Martinez to collect her payment—in gold.

ORINDA PARK SCHOOL. Located near where Wagner Ranch Elementary now stands, Orinda Park School (pictured around 1920) was built in 1882 on land donated by Gen. Theodore Wagner. On stormy days, when roads were impassable, a red flag would alert students to a "no school" day. In an era of long and wet winters, these impromptu vacations could easily last for a month or two. Arson was suspected when the school burned down in 1885, but it was soon rebuilt and continued to be used until 1924, when enrollment, which was initially 14 students, had grown too numerous for all 10 grades to be taught in a one-room schoolhouse. The building was auctioned off in 1925 for $62.50. Note the number of stars on the flag in the below photograph, which was taken around 1898.

1st ORINDA UNION SCHOOL. 1925 2 Rooms + Auditorium.

ORINDA UNION SCHOOL. By 1924, Orinda's students had outgrown the 19th-century one-room schoolhouses. The following year, the Orinda Park and Moraga School Districts, renamed the Orinda Union School District, opened the two-classroom Orinda Union School. March 7, 1925, was declared "tree planting day" (above), with impressive results. By 1939, when the first- and second-graders posed for the below photograph, extensive rebuilding (note the rubble at their feet) was necessary to fix the school's deteriorating materials and to increase the number of classrooms—this work was carried out with the support of the Public Works Administration. Elementary students were taught at the Orinda Union School until 1974, when the building was sold and reconfigured as the Orinda Community Center.

TWENTIETH-CENTURY SCHOOLS. In 1947, the entire Orinda Union School student population could still fit on the bare hill behind the school, but it was obvious that more schools were urgently needed to deal with the town's rapid growth. Four elementary schools and one high school opened between 1949 and 1961, and by 1969, when Wagner Ranch (below) appeared, Orinda's school roster had reached its peak. Then the decline began—three schools closed in 1975, followed by Wagner Ranch Elementary School in 1982. However, as new families discovered Orinda, schools were again needed, and Wagner Ranch, thoroughly updated, reopened in 1997. (Below, CCCHS.)

GLORIETTA ELEMENTARY. As the baby-boomer generation approached school age, the Orinda Union School District realized that the old Union School on Orinda Way could no longer accommodate the burgeoning influx of postwar children. The first new elementary school, Glorietta, opened in 1949 down the road from where Orinda's first school had been built in 1861. (Orinda Union School District.)

SLEEPY HOLLOW ELEMENTARY. Although developers arrived in Sleepy Hollow in the 1920s, it was only in the postwar years that young families arrived en masse, eager to raise their children in this sunny corner of Orinda. Built in 1953, Sleepy Hollow Elementary now counts among its students the grandchildren of some of the area's early families. (Orinda Union School District.)

DEL REY ELEMENTARY. This photograph shows Paul Barrett lining up to play baseball in 1954; he was part of the first intake in the newly built Del Rey Elementary. Paul's older brother Reg was designated the school's official photographer, largely due to the fact that he owned a Kodak Brownie camera. (Orinda Union School District.)

INLAND VALLEY INTERMEDIATE. Inland Valley Elementary and Inland Valley Intermediate Schools opened in 1960 and 1961, but like Pine Grove Intermediate and Orinda Union School, they suffered from falling enrollment in the 1970s. When the two Inland Valley schools closed in 1975, the Inland Valley campus morphed into a single merged school now known as Orinda Intermediate. (Orinda Union School District.)

A Historic Oasis. Oak Springs Pool, pictured in 1926, started in 1900 as a man-made swimming hole fed year-round by mountain spring water. Legend has it that the pool was actually an immense redwood wine vat sunk into the ground. The pool was rebuilt in 1929 by the Oak Springs Home Owners Association, and a clubhouse was added in the 1940s.

The Adobe Pool. According to William White III, whose family owned the Moraga Adobe from 1912 to 1965, their pool, pictured in 1966 with the Manuel family, boasted "the springiest diving board." It was built around 1954 by handyman Hugh Pinkley with a little help from his sons. Pinkley had never constructed a pool before but "just figured it out, like a horse trough." (Moraga Historical Society.)

ORINDA PARK POOL. Orinda Park Pool began as Lake Orinda in the early 1920s as a water supply for E.I. de Laveaga's first subdivision, and it later became a pool and picnic area. It was renovated in 1931 with a dance platform and opened to the public. However, nearby residents grew unhappy with so many raucous strangers on their property and formed the nonprofit Orinda Park Pool, Inc., as a private association in May 1938. For many years, coach Peg Kirby was a much-loved fixture at the Orinda Park Pool, and many Orindans still cherish fond memories of riding around the pool in her pony cart. (Right, Kirby family.)

SLEEPY HOLLOW. In March 1955, members of the newly formed Sleepy Hollow Recreational Association met to discuss plans for the 55 acres they had recently acquired for below market value—plans that included an impressive swimming pool, a clubhouse, a children's wading pool, and a picnic area. Work began in June of that year, and while all of the major construction was contracted out, association members willingly contributed thousands of hours of volunteer labor, often in scorching heat, in order to keep costs down. Those who were not able to spare time were encouraged to contribute $10. By the spring of 1956, membership (which cost $240) was sold out, and by June of that year, Sleepy Hollow's pool was officially open. (Both, Sleepy Hollow Recreational Association.)

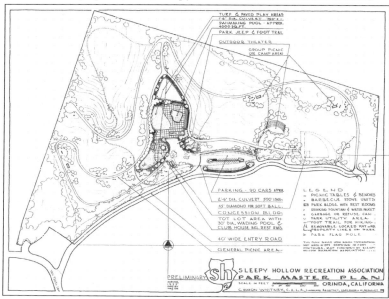

ORINDA COUNTRY CLUB POOL. According to the elegant 1924 brochure aimed at attracting buyers to the "congenial community" of Haciendas de Orinda, the free-form pool at the Orinda Country Club (OCC) had sides "hid in growths of shrub and vine and flower" and was modeled on the "Blue Hole" pool at the Mokelumne River. It was fed from Lake Cascade via a 35-foot man-made waterfall, and its varying depths were said to make it as safe for small children as for adults. The bathhouse, built in the OCC's trademark Spanish architectural style and presented in the brochure as a pastel sketch, was coyly promised to be "complete in all its appointments."

A Big Thumbprint. The dominant physical process shaping Lamorinda's topography is undoubtedly its landslides. In December 1950, a massive slide "like a big thumb print" dumped 1 million tons of rock, earth, and other debris on Highway 24, blocking all four lanes. It was not possible to start clearing it until the following April—and it was an exercise that continued well into the summer of 1951. The below photograph gives a sense of the enormity of the task merely by noting the body language of the men who are craning their necks to take in the overwhelming problem they were asked to solve.

Six

Earth, Wind, and Fire

Records show that Orinda has experienced over 4,500 earthquakes since 1931. So far, they have tended to be more inconvenient than deadly. But Orinda has also suffered its fair share of mudslides.

In the early 1970s, a geological survey undertaken by the US Department of the Interior concluded that slides like the ones Orinda experiences from time to time largely occur on preexisting ancient landslide deposits as a result of grading or construction. During the El Niño winter of 1997–1998, over 100 landslides occurred within Contra Costa County.

Floods also occur in Orinda—during the Columbus Day storm of 1962, only one bore of the Caldecott Tunnel remained open for just a few hours a day after cascading mud and water knocked out its electrical generators.

Snow is extremely rare. Longtime residents may remember flurries in 1976, and perhaps there are a few who still recall making a snowman in the winters of 1937 or 1955. Such extraordinary memories seldom fade.

But not everything falling from above can be blamed on Mother Nature. In the 1950s and 1960s, once the results of America's first H-bomb test were publicized, the population learned to live in fear of radioactive fallout from enemy bombing. If you should venture to a website dedicated to people who grew up during that era, you might be surprised at how many remember their parents following Pres. John F. Kennedy's heroic instructions to build a backyard bomb shelter, complete with spiders and canned goods.

MEMORABLE WINTERS. March 1969 proved particularly difficult for the owners of this house when their backyard unexpectedly disappeared. Even now, Orinda continues to be vulnerable to heavy rains; January and February 2017 witnessed the Miner Road sinkhole, a hillside that crashed into a Sleepy Hollow home, and a PG&E tower that threatened to topple when a mudslide rearranged the nearby landscape. The Columbus Day storm of October 1962, ranked third for severity in the Bay Area Storm Index, registered winds of 43 miles per hour and gusts of 86 miles per hour. The below photograph, taken at the old Orinda Crossroads, demonstrates the patience of one Orindan prepared to sit it out on the hood of his car. (Above, CCCHS.)

THE NURSERY. Snow was (and is) so rare in Orinda that the moment it appears, the cameras come out. Those photographs allow the world to catch a glimpse of Orinda Way in 1937. The Orinda Nursery, built in 1926, was located next to where Orinda's Christian Scientists held their first meeting in 1935.

FIRE AND SNOW. In 1935, two years before this photograph was taken, a small library was tacked onto the side of the newly remodeled firehouse, and a librarian was employed for $10 per month (reduced to $7.50 during the Depression). After the fire department moved out in 1942, the library was granted a short reprieve but had to relocate when the building was sold in 1944.

ONE HUNDRED YEARS OF FIREHOUSES. A volunteer fire department was one of E.I. de Laveaga's earliest priorities, and in 1923, a firehouse just large enough to accommodate one truck was constructed on Orinda Way. It was remodeled and enlarged in 1935 but was soon considered too small, and new premises were built in 1942. This also had its problems, and in 1969, the fire

department moved to its current location on Orinda Way. The original firehouse is still standing and, over the years, has housed a real estate office, a furniture store, a hair salon, an art gallery, and both Harnett Antiques and Orinda Village Antiques.

THE JUNIOR FIRE DEPARTMENT. In 1939, fire chief Joe Varni invited boys under 15 years old (boys only) to join his newly formed junior fire department. While it was a fun experience, the boys also learned the practicalities and self-discipline that would place them in good stead during the upcoming war years. Many of the juniors went on to become volunteer firemen. Their regular drills included identifying hazardous situations within the community, which was very popular. However, nothing proved quite as popular as setting fire to old sheds and extinguishing it before it got out of control.

THE 1942 FIREHOUSE. At the end of Avenida de Orinda stand two metal shafts set in concrete—the last reminder of the flagpole that once sat outside this handsome brick firehouse. Located on a dead-end street close to San Pablo Creek, the firehouse, which is pictured in 1945, was vacated in 1969, when the fire department relocated to Orinda Way to avoid the creek's regular flooding. The abandoned firehouse was demolished in 1972, much to the chagrin of local high school students who had grown to like their secret meeting place. Fire chief Al Winsor is pictured below outside the firehouse in 1959. (Below, CCCHS.)

BEAR CREEK RESIDENT. It is fitting that this injured bear was picked up by a California Highway Patrol officer on the aptly named Bear Creek Road. According to local vertebrate zoologist and wildlife biologist Dr. James Hale, grizzlies and black bears roamed the area, drinking from the San Pablo Creek, until the mid-1960s.

THE BRIDGE INTO ORINDA. After the San Pablo Dam was built in 1918, the road into Orinda suffered severe flooding whenever the creek was high, making access impossible. Eventually, the old wooden bridge (pictured in 1890) was replaced with something more substantial, which still straddles the San Pablo Creek more than 100 years later.

THE PELLISSIER FALLOUT SHELTER. When the Cold War began to heat up, a letter from Pres. John F. Kennedy appeared in the September 15, 1961, issue of *Life* magazine suggesting that his fellow Americans might like to start building their own fallout shelters in preparation for a potential nuclear war. As long-term sustenance, the government would supply biscuits made out of bulgur wheat, which apparently was the ideal food, since edible bulgur had recently been discovered in a 3,000-year-old Egyptian pyramid. The Department of Defense issued instructions for eight different homemade shelter designs. Far more appealing, however, at least in Orinda, was Pierre Pellissier's luxurious version, which—thankfully—was never needed. It saw better service as a children's playhouse and even outlasted the Berlin Wall. (Both, CCCHS.)

DRILLING FOR GOLD. In the late 1880s, oil hysteria in Orinda led to a proliferation of wells, particularly around Miner Road. All of them were doomed to fail. However, after many years of drilling on his ranch in present-day Sleepy Hollow, James Miner finally struck it rich. His first attempt had produced five gallons, which he slowly fished up with a bottle, but he was unable to continue when his drill broke loose. He tried again in 1903, and this time, the oil "shot up into the air." Unfortunately, it hit some nearby lanterns and started a huge fire that destroyed all his machinery and almost killed his workers. In 1939, the Orinda Petroleum Company, which had been drilling near Upper El Toyonal Road, admitted defeat and allowed its 122-foot derrick to be dynamited and removed. Orinda was never going to be Houston.

Seven

GOING OUT, STAYING IN

In less than a decade, E.I. de Laveaga had created a fully functioning village far from foggy San Francisco and a whole new lifestyle to tempt California's nouveau riche. Around that time, few people commuted on a daily basis, so building lots were largely aimed at those able to afford a second home. By 1929, the Roaring Twenties had given way to the Great Depression, and within four years, half of the country's banks had failed, leaving 15 million Americans out of work.

By the end of World War II, the United States enjoyed better economic conditions than any other country. Returning GIs were looking for safe suburban communities with temperate climates, good schools for their kids, and an easy commute to the city. Orinda ticked all of these boxes. Restaurants flourished again, and leisure time could be spent watching the latest movie in the new Orinda Theatre, shopping at stores where everybody knew your name, or just chilling with friends in one of the new community pools.

Even Frank Lloyd Wright, who designed fewer than 20 houses in California, was persuaded to build one of his revolutionary Usonian homes in Orinda. The owners were happy to shut the world away and stay in, admiring their home's Zen qualities, but when they left for any length of time, such as a round-the-world trip in 1971, they carefully positioned several buckets under their exquisite yet leak-prone roof.

THE ORINDA THEATRE. On December 27, 1941, while the country was still reeling from the bombing of Pearl Harbor, the Orinda Theatre opened its doors to the public with a double feature of *The Maltese Falcon* and *Tarzan's Secret Treasure*. The Orinda was designed by Alexander Cantin, who was eventually responsible for 39 exquisite movie theaters, 10 of which are still open. The theater was constructed in a stripped down, late Art Deco style that incorporated curved forms,

Carrara glass, and plenty of neon. The theater's terraced pyramid exterior echoes the Court of the Moon portals at the 1939–1940 Golden Gate International Exposition, while the interior was designed by muralist Anthony Heinsbergen, who created over 750 theater murals, including Oakland's Paramount theater, and covered the walls and ceiling in fantastical mythical figures representing earth, fire, air, and water.

CONSTRUCTION OF THE ORINDA THEATRE, 1941. This photograph shows the theater's construction in 1941, by which time its architect, Alexander Cantin (1874–1964) had spent 20 years putting his increasingly famous name to movie theaters throughout the Bay Area. In 1948, he renamed his practice Cantin & Cantin when his son, Mackenzie, joined him and together they designed the American Trust Building, show at far left in the picture below and on page 106.

SAVING THE ORINDA THEATRE. This photograph dates from 1953, when Hitchcock's film noir *I Confess* was still pulling in the crowds. In 1982, local realtor Clark Wallace bought the four-acre block, intending to tear down this "ugly monolithic mausoleum" and replace it with a five-story, 116,000-square-foot building containing three mini-theaters. Historical and preservation societies opposed the plans, the Orinda Association cited height and density issues, and the Friends of the Orinda Theatre presented a 22,000-signature petition. Even the California Supreme Court got involved. Eventually, a compromise was reached that resulted in 35,000 square feet of retail space, 40,000 square feet of offices, and a three-screen theater that opened on June 29, 1989, with a showing of *Casablanca*.

RHEEM'S DREAM. Although Donald Rheem's name is largely synonymous with Moraga, this passionate cinephile (pictured) forever altered Orinda's landscape in 1941, when he spent $400,000 to construct the town's most beloved landmark, the Orinda Theatre—even though the golden age of "popcorn palaces" was about to decline as television sales rose. In 1983, shortly before Rheem died, when he was told that his beloved theater was in danger of succumbing to the wrecking ball, he replied, "that's progress." Rheem would be pleased to know that more than 80 years after that dorsal fin first pierced the Orinda sky, it continues to shine like a welcoming beacon in the night.

RITE AID RODEO. For 15 years, beginning in the mid-1920s, Edgar Ingram used his photographic talents to produce artwork for real estate brochures and everyday events. In 1939, when he was packing up his papers to become a Berkeley antiques dealer, Ingram scribbled these words in pencil on a scrap of paper: "On some far distant day someone will discover these old catalogues etc etc and stir their interest about 'the good old days.' I wonder how long before these things are discovered?" More than 80 years later, his rediscovered photographs document an entirely different time and place, when massive crowds gathered in the 1930s for Orinda's popular rodeo held on what became the Rite Aid parking lot.

ORINDA COUNTRY CLUB. When E.I. de Laveaga planned Haciendas del Orinda, he knew he had to pull out all the stops to entice potential buyers to make the long and often treacherous journey to Orinda. So, he built the perfect country club (above) at a cost of almost $600,000, with membership available to anyone who purchased one of his homesites. In 1924, a year before the club's formal opening, regular membership was established at $500 per year with $10 monthly dues. Anyone opting for life membership in 1925 had to part with $1,000 but avoided monthly dues. The photograph below shows the completed clubhouse by Lake Cascade. The house in the foreground, located just off Miner Road, is Casa Azul, constructed in 1924–1925. Note the PG&E towers already straddling the countryside.

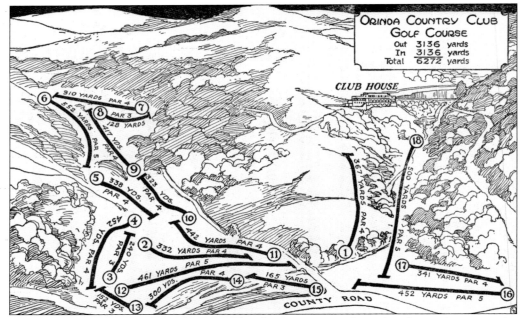

ORINDA COUNTRY CLUB GOLF COURSE. This diagram appeared in a brochure advertising the Haciendas del Orinda subdivision. According to Ned de Laveaga, his father, E.I, had "never touched a golf club." But he knew he wanted the world's best golf architect, whom he found in William Watson. Watson insisted that the first nine holes should end at the clubhouse, with the 10th hole nearby, and E.I. was adamant that his clubhouse must be situated where guests could sit and gaze across the San Pablo Valley after a game. After much arguing, Watson relented—this explains why even now, golfers playing nine holes must skip from the 3rd to the 13th hole in order to end at the clubhouse by the 18th green (which is pictured below in 1924).

Miss Graham's Riding Class. Mrs. Philip Donaldson, proprietor of Miss Graham's Riding Academy on Orinda Way, poses in front of her class in front of the Orinda Country Club in 1925. Third from the right, on the horse with the long white forelock, sits Aloyse Goelitz, whose father, Herman, owned the Herman Goelitz Candy Company, makers of candy corn and (after 1960) Jelly Belly jelly beans.

Alaska or Bust. In 1959, just months after Alaska became a state, Miramonte High School seniors Chad Dressler and Marc Littlejohn embarked on a 6,000-mile round trip to the 49th state in their Model A Ford. Many years later, their former classmates still recall the two teenagers as "legends among us younger guys." (CCCHS.)

101

THE WILLOWS. These photographs were taken in the mid-1950s, when The Willows restaurant was the place to spend an evening or hold one's wedding reception. Guests could dance to Wally White's "Tip Tops" and enjoy a sumptuous five-course dinner for two—with wine—for just $10, while an in-house photographer locked the occasion into permanence. Various restaurants had been on this site since 1917. The Willows, which opened in 1938, was named for the trees growing nearby—not to mention the one shooting up through the floor and out the roof. (Below, Janet Ish Bennett.)

GUNS AND PRIME RIB. In 1932, one could choose from four eateries at the Orinda Crossroads—one on each corner. Three of them are long gone, but Casa Orinda, which moved to its current location in 1942, now ranks as Contra Costa County's oldest continuously operating restaurant. Cavernous and windowless, the Casa has changed little in its 90-plus years: the decor is still cowboy-themed, with antique guns displayed on the walls and fried chicken and prime rib on the menu. In the fireplace room, the tables remind 21st-century diners of those early days. When the 1930s hand-carved mahogany bar was installed, Casa Orinda's owner, Jack Snow, invited his rancher friends to bring over their branding irons and brand all the tables.

CANARY COTTAGE. Canary Cottage, pictured in 1925 under Orinda's first snowfall in eight years and its heaviest since 1882, was popular with motorists and teamsters exiting the unlit Kennedy Tunnel. Originally called the Tunnel Inn, the roadhouse was renamed Canary Cottage after its owner, Frank Enos, persuaded the Shell Oil Company to paint it yellow. The journey through the tunnel improved slightly after the power company installed electric lights in 1919, at which point Frank offered to replace any burned-out lights in return for free electricity. That was the last winter Frank and Ruth Enos spent in Canary Cottage before relocating to their new restaurant and gas station on Orinda Way. This was Orinda's first commercial building, and the couple again chose an ornithological name, calling it the White Swan, now known as Casa Verana. (Below, author's collection.)

ORINDA PARK HOTEL. In 1885, Theodore Wagner built this hotel to the north of Bear Creek Road in anticipation of the extension of the California & Nevada Railroad. The hotel always struggled, mostly because since it was close to Orinda Park School, it was impossible for the hotel to obtain a liquor license. Once the railroad failed around 1900, Wagner cut his losses and sold. For a while, the hotel was used as an invitation-only social club (below), but it was torn down in 1913. Although the hotel is gone, the original foundation remains just outside Orinda city limits on East Bay Municipal Urban District property. The above photograph shows the hotel around 1907, after it had fallen into disrepair.

Orinda Park Social Club.

Admit Mr. .... and Ladies,

Saturday Evening, ... 189 ......

Compliments of ....

ADMISSION, 50 CENTS.

THE X ROADS. The Crossroads Restaurant opened in 1928 next to Orinda's first Standard Oil gas station, and the restaurant relocated west of the new Highway 24 in 1932. Don and Ruth Thompson bought the restaurant in 1945 and turned it into one of Orinda's finest dining establishments. Sadly, like so much else in 1950s Orinda, it ended up as one more casualty of the highway-widening project.

AMERICAN TRUST COMPANY BANK. Orinda's first bank opened in 1947 and was designed by the same father-son team, Cantin & Cantin, that had produced the theater next door. In 1960, the American Trust Company Bank merged with Wells Fargo to become the Wells Fargo Bank American Trust Company; however, within two years, it was known simply as Wells Fargo Bank. (CCCHS.)

PHAIR'S. Mention the name Phair's to anyone who grew up in Orinda, and they will immediately enthuse over the specialty shop on Avenida de Orinda that, for almost 60 years, supplied gourmet foods, fine wines, wonderful dresses, and best of all, wedding gifts. When it first opened in 1925, it was simply known as The Store, offering hardware supplies, a beauty shop, a post office, a candy store, a soda fountain, and even a fledgling library, which consisted of two rows of books in a five-foot bookcase. After Ewart Phair took over in 1941, it grew into the legendary Phair's, which closed down in the late 1990s and has remained empty ever since. However, its latest owner, Orinda native Joanna Guidotti, plans to breathe new life into the legendary store and make it the indispensable hub it once was. (Below, author's collection.)

BLACK'S MARKET. Opened in 1947 by Frank Black, Black's Market was an Orinda institution for over 40 years and was renowned for its high quality and attentive service. The market was located at 6 Camino Pablo, where Bev Mo now stands. In 1960, Italian-born Vasco Giannini—pictured at far left below in 1948 with, from left to right, Tony Cardinali, John Hewitson, and Vince and Frank Maita—bought Black's Market and opened a second store (now a Rite Aid) in Orinda Village, plus a liquor store in Moraga. Giannini was a larger-than-life character who exemplified community spirit and dependability. After he retired in 1986, the store changed hands twice more before being sold to Payless in 1991.

MOVING DAY. Bradley's Cash Market opened in 1943 at the old Orinda Crossroads and remained there until the lease expired in 1951. After the new owners moved the building to Brookwood Road (via Highway 24), they traded for another two more years until the road-widening project began. Those perfectly poised men atop the building are there to raise cables as necessary.

BILL'S PHARMACY. Before it was CVS or Longs, the familiar pharmacy at the corner of Moraga Way and Brookwood was Bill's Pharmacy, part of a 20-store chain of drugstores in Northern California. This 1983 photograph, taken 10 years before Longs paid almost $24 million to acquire Bill's, shows how significantly Orinda's landscape has been altered.

ORINDA MOTORS. Miss Graham's Riding Academy on Orinda Way is long gone, but the garage next to it, which was built by E.I. de Laveaga in 1924, remains. It is now home to Orinda Motors, which, since 2005, has raised substantial funds for local charities by hosting the city's annual Classic Car Show. The 1920s cast-iron lamppost in the foreground of the below photograph was rescued by local artist Joe Cleary in the 1970s and stood outside Joe's home on Tarry Lane for almost 50 years. After Cleary died in 2019, the new homeowner, Dr. Chiyo Shidara, happily assumed guardianship of Orinda's 17th historical landmark. (Above, Kent Long.)

PLANTING PINES. In the spring of 1952, one of Orinda's civic planting projects consisted of placing hundreds of pine seedlings around the Orinda Community Church and on community center property. At the time, Orinda was growing houses faster than forests, and many people felt the landscape needed softening. Monterey pines, which could be grown cheaply from tiny saplings, were the answer. The Girl and Boy Scout troops (Pack 32) joined in with alacrity, with few of them knowing just how fast their saplings would grow or that 70 years later, bark beetle and droughts would turn the pines into potential fire hazards. (Both, CCCHS.)

BOY SCOUTS AND GIRL SCOUTS. Orinda's first Boy Scout troop was launched in 1933, but since it was one Scout short of qualifying for sponsorship, a local family offered their Chinese house boy, Wing Wong Quong, as a charter member. In 1936, the Girl Scouts began their own program under the umbrella of the Parent Teacher Association.

CAMPING OUT. In 1896, when the unidentified photographer created this peaceful family tableau, Orinda was still known as Orinda Park, the name that William and Alice Camron had given to the 3,000 acres they purchased in 1875. The word "Park" was dropped at the beginning of the 20th century. Note the woman in the hammock strumming a banjo.

WILD CAT CANYON, 1897. If anyone should wonder why Wild Cat Canyon was so named, they have only to look at this photograph and the animal strung up on the tent pole while four dogs rest beneath the buggy like sentinels ready to challenge all interlopers. Take a close look at the gentleman seated at far right—is that a raccoon sitting on his knee?

TRAVELING HORSESHOER. It was a sunny Wednesday in August 1959 when traveling horseshoer Bill Hardy visited Jeanne Patten and her horse, Sunday, in Orinda. A good farrier, it is said, needs to combine the skills of a blacksmith with those of a veterinarian (although in Britain, when the queen's farriers march in parades, they are also required to wear ceremonial dress and carry historical axes with spikes). (CCCHS.)

WAGNER RANCH NATURE AREA. The public gets to visit the Wagner Ranch Nature Area once a year for Orinda's annual Olive Festival. But Orinda's schoolchildren are allowed to go more often—since 1970, thousands of students have enjoyed a unique, hands-on environmental education as they explore what was once Gen. Theodore Wagner's Oak View Ranch. Featuring 18 acres of meadow, forest, ponds, and streams, the Wagner Ranch Nature Area, which is bordered on one side by the San Pablo Creek (pictured above in 1910), is home to thousands of native plants. As shown in the 2008 picture below, Richard Adams (center) accompanied his wife, Pat (left), a direct descendant of General Wagner, on a visit to the historic area. On hand to tell Wagner's great-granddaughter about all the wonderful work accomplished there was Toris Jaeger (right), who has worked at the nature area for over 45 years and has had an immeasurable impact on the young minds she constantly inspires.

ORINDA'S FRANK LLOYD WRIGHT HOUSE. In 1948, Katie and Maynard Buehler asked Frank Lloyd Wright to design them a house in Glorietta. The result was a magnificent 4,000-square-foot Usonian home set on 3.5 acres with gardens created by Henry Matsutani—including a fish pond that contained koi flown in from Japan in a bowl of water on Katie's lap. Under the distinctive, cantilevered copper roof sits a unique octagonal living room that slopes from 2 feet to 14, its ceiling inset with 22-carat gold leaf to reflect light into the space below. After the project was completed, Wright visited the Buehlers, poking various objects as he walked around, and finally commented, "I'm glad to see you're living graciously in my house." (Both, author's collection.)

THE OLD YELLOW HOUSE. In 1918, with a flu pandemic raging across the world, Charles Nelson moved his young family from Albany to Orinda and settled into the Old Yellow House on Moraga Way. After Ezra, the youngest Nelson, moved out in 1966, the house remained empty for almost 50 years until architect James Wright converted the 1894 structure into a net-zero-energy building. Despite it now being solidly futuristic, with a geosolar envelope that ensures Wright's home is heated by the sun and cooled by the earth, the house still conveys a strong sense of history.

THE OLD YELLOW HOUSE HALLWAY. In the hallway, perfectly preserved newspapers, once used as insulation under ancient linoleum during the Depression and World War II years, have been lovingly repurposed as wallpaper lining the entire staircase and evoking a long-forgotten era. (Author's collection.)

DINING WELL. In the basement of the Old Yellow House, a round, glass-topped dining table designed by James Wright reveals a 27-foot sheer drop down a brick-lined well, which Charles Nelson dug by hand more than 100 years ago. Nelson earned $5 a month hauling water to the nearby 1861 schoolhouse until plumbing was installed there in 1924. (James Wright.)

MOTORBIKES. Earl Nelson, pictured in 1928 with his younger brother Ezra seated behind him, recalled how his father once drove him to an Oakland dentist on his Thor motorcycle with young Earl perched up front on the fuel tank. Before starting out on a journey, the rider would light the bike's gas headlamp with a wad of burning paper so they could see their way through the pitch-dark 1903 Kennedy Tunnel.

HANDSOME 12-ROOM RESIDENCE
ORINDA, CALIFORNIA

Pine Crest Manor, a well-built Mediterranean home in the heart of Orinda Valley, lies on a gently sloping hillside paralleling the first fairway of the Orinda Country Club. The wall and fence enclosed 3.2-acre grounds are well landscaped with terraced lawns, shrubs and trimmed hedges. Towering pines and deodars, as well as a lower garden with pool and statuette, add to the charm of the setting.

The 12-room house is reached along a curving driveway leading from the pillared entrance gates, past the 4-car garage and servants' apartment to the front entrance, and thence to the covered port-cochere. The main residence features a spacious living room with high curved ceiling and fireplace, cheerful library with fireplace and built-in book cases, and dining room with exquisite Czechoslovakian crystal light fixture. A glassed, tile-floored solarium and outdoor patio take full advantage of a climate delightful the year around. All 2nd floor bedrooms are of good size and each has its own bath and shower. A 3rd-floor tower room, bright and cheery, can be a children's playroom.

A fine home in a choice country setting. Pine Crest Manor has the added attraction of being only 18 minutes' drive from Oakland, 20 miles from San Francisco.

LOCATION: Miner Road, Orinda, Contra Costa County, California, an area of attractive homes. Adjacent to 1st fairway of Orinda Country Club. School bus at gate, shopping center ¾ mile, Oakland, 9 miles, San Francisco 20 miles.

GROUNDS: Approx. 3.2 acres, wall and fence enclosed, pillared entrance gate, curving driveways. Landscaped with trimmed hedges, terraced lawns, shrubs, large trees, deodars, pines and others, planting beds, formal rose garden, pool with statuette, and family orchard.

RESIDENCE: 12 rooms (4 master bedrooms, 4 baths, 2 servants' bedrooms and bath). Two-story, built in 1932, redecorated 1950, painted recently. Frame and stucco construction, concrete foundation, tile roofs, hardwood flooring throughout. Central heating (steam) thermostat controlled, hot water heater attached to furnace. Town water. Public utilities. Town sewerage.

MAIN FLOOR: *Entrance Hall* (15 ft. x 9 ft.), coat closet and iron-railed stairway to left. Also *Living Room* (20 ft. x 30 ft.), curved ceiling, fireplace with Italian marble mantle, windows overlooking garden. *Library* with fireplace. Beyond is *Solarium* (14 ft. x 7 ft.) opening to outdoor patio. *Dining Room* (21 ft. x 16 ft.) with overhead Czechoslovakian crystal light fixture. *Pantry* with sink, *Kitchen* (17 ft. x 10 ft.). Service porch, back hall, 2 servants' bedrooms with connecting bath.

2nd FLOOR: Long Hall. *Bedroom* (19 ft. x 13 ft.) with sitting room. *Bedroom* (17 ft. x 14 ft.). *Master Bedroom* (19 ft. x 15 ft.). French doors to covered balcony dressing room. *Bedroom* (20 ft. x 14 ft.). Each bedroom has its own tile bath. Two baths have stall showers, two have tub showers.

3rd FLOOR: *Tower Room. Basement:* Partial.

GARAGE: Detached 4-car. Bedroom and bath, 1st floor. 2nd floor living room, bedroom, kitchen and bath.

ADDITIONAL BUILDINGS: Play House (running water), lath house.

MORTGAGE: $25,000 . . . TAXES: $1,260

OFFERED AT: $82,500 Unfurnished. Terms may be arranged.
Included with sale are drapes, hall and stair rugs. Additional furnishings may be purchased from owner.

PREVIEWS LISTING No. 70517

Offering is subject to errors, omissions, prior sale, change, withdrawal without notice, and approval of purchaser by owner.

LIVING ROOM

DINING ROOM

LIBRARY AND SOLARIUM

OWNER: J. P. French, 40 Miner Road, Orinda, California. Telephone: Orinda 2240.

PROPERTY: Pine Crest Manor, 40 Miner Road, Orinda, California.

INSPECTION: By appointment with owner on property, or Previews Inc.

DRIVING DIRECTIONS: From Oakland out Broadway to Orinda crossroads 9 miles. Turn left past Orinda Village to Miner Road 1 mile. Turn right. Residence 1st house beyond Catholic Church on right.

No. 70517   SF 553
Owner authorizes one commission up to 5% of selling price to selling broker.

PREVIEWS INCORPORATED, *The National Real Estate Clearing House*
68 Post Street, San Francisco 4, California, DOuglas 2-3006

New York   Boston   Philadelphia   Palm Beach   Chicago   Denver   Los Angeles

CASA AZUL. Another stately 1920s home still cherished today is the elegantly graceful Casa Azul. In 1925, when Chester Williams II took over the home his father had commissioned but never lived to see, 40 Miner Road was an unnamed Spanish Mediterranean house finished in drab tan stucco. Twenty years later, Chester sold the house to Mr. Salisbury, a plumbing contractor who, five years later, sold it to Mr. French, a Safeway executive. At some point, the house, still tan in color, became Pine Crest Manor. In 1957, Mr. French listed it at $82,500 but settled for $60,000 when the Calhouns moved in with their seven children. By that time, the house was beginning to show its age, so Bill Calhoun painted it blue and renamed it Casa Azul—which it has been known as ever since. In 1964, the eight-member Garbarino family bought Casa Azul and held onto it until 2020, when it was bought by a family who are happily restoring this stunning Orinda landmark to its former glory. (D'Anna family.)

THE PLAYHOUSE. According to a 2001 interview with Joyce Williams Hart, who was born in 1933 and was the first child to live in Casa Azul, her father, Chester Williams II, was a "party type" who never had to work after inheriting substantial wealth from his father. As an only child, Joyce, pictured above with her mother, Adelaide Robbins Williams, and grandmother, grew up with a French nanny who lived in the "tower room." In 1938, when Joyce was five, Chester built his daughter a playhouse with miniature windows and doors all to scale, but sadly, Joyce had to leave Casa Azul in 1943, when her parents divorced. Today, toys are again scattered outside the playhouse as the sound of children's laughter echoes from inside. (Both, D'Anna family.)

THE BROBECK LAWN MOWER. In the mid-1950s, the Brobeck family's Sleepy Hollow backyard proved a source of endless fascination as neighbors looked on in awe, unaware that Bill Brobeck Sr. had buried electrical wires beneath the lawn to guide his latest invention, an automatic mower. Housed in a nearby shed, it would start itself up and head out to the lawn, cutting alternate swaths while moving ever closer to the center, then turn around and cut the remaining grass on the return trip, while young Billy Brobeck (pictured) kept his distance. When the entire lawn was done, the mower would head back to the shed, switch itself off, and recharge the battery. Bill Brobeck is pictured above with the mower. (Both, Bill Brobeck.)

PHONE HOME. This eight-page Orinda telephone directory, which was issued in 1938, the year Orinda got its own exchange, contains just 225 names—something that some 21st-century Orindans should find fascinating, since it reveals who lived at their addresses in 1938. The directory also gives the following advice on dealing with this novel apparatus: "It is not necessary to shout—talk directly into the telephone with your lips just clearing the mouthpiece and in an even tone of voice to insure that your telephone conversation will be heard satisfactorily." By 1940, underground telephone wires were being laid in Orinda Village, and by 1957, the tally of phones in town had risen to 5,600.

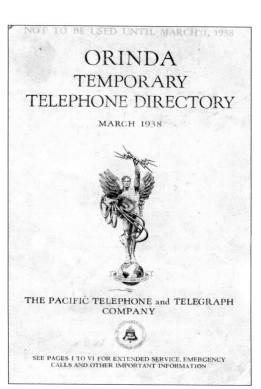

ORINDA
TEMPORARY
TELEPHONE DIRECTORY

MARCH 1938

THE PACIFIC TELEPHONE and TELEGRAPH
COMPANY

SEE PAGES I TO VI FOR EXTENDED SERVICE, EMERGENCY
CALLS AND OTHER IMPORTANT INFORMATION

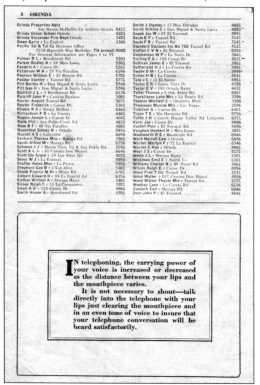

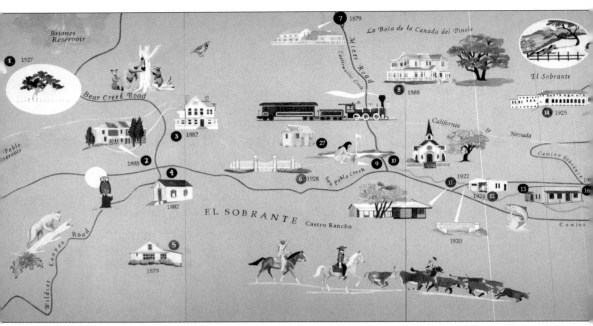

BICENTENNIAL MURAL. In 1976, in honor of the 200th birthday of the United States, the Orinda Historical Society commissioned local artist Lonie Bee to create a mural celebrating many of the town's historic places. The 23-by-6-foot painting hung at the BART station for 30 years, but despite the efforts of local artists to repair it in 1997, it became increasingly fragile over time. In 2011, muralist Ellen Silva recreated Bee's illustration, adding two more historic landmarks—the

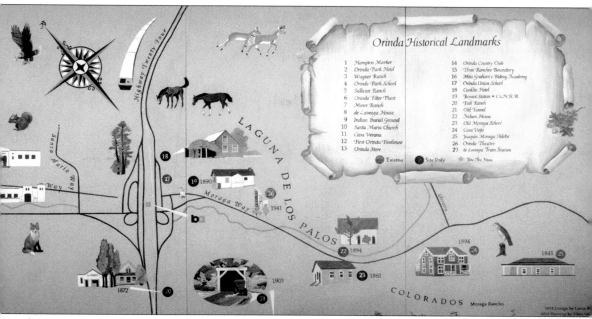

Orinda Historical Landmarks

1	Hampton Marker	14	Orinda Country Club
2	Orinda Park Hotel	15	Three Ranches Boundary
3	Wagner Ranch	16	Miss Graham's Riding Academy
4	Orinda Park School	17	Orinda Union School
5	Sullivan Ranch	18	Conklin Hotel
6	Orinda Filter Plant	19	Bryant Station • C&N.R.R.
7	Miner Ranch	20	Fish Ranch
8	de Laveaga House	21	Old Tunnel
9	Indian Burial Ground	22	Nelson House
10	Santa Maria Church	23	Old Moraga School
11	Casa Verana	24	Casa Vieja
12	First Orinda Firehouse	25	Joaquin Moraga Adobe
13	Orinda Store	26	Orinda Theatre
		27	de Laveaga Train Station

de Laveaga train station and Orinda Theatre. The following year, Silva's bright new replica was installed on the KinderGym wall at the Orinda Community Center Park. Coincidentally, this photograph was produced by someone born on the Fourth of July in the year of the Bicentennial. (Sean Slinsky.)

GUNS IN SCHOOLS. The days when students could walk into Orinda's schools carrying a gun are ancient history, but in 1954, when Reg Barrett proudly earned his NRA badge, firearm safety was still being taught in local schools. Miramonte High School rifle team members from the 1960s and 1970s recall having their guns locked away first thing in the morning and then practicing at the Chabot range in the afternoon.

BOBBIE LANDERS. Orinda's first female mayor, Bobbie Landers, is pictured in the driver's seat in the mid-1990s alongside former mayor Aldo Guidotti, while Linda Knebel, along with former mayors Bill Dabel (left) and Dick Heggie (right), waves from the backseat. Landers has been Orinda's driving force for decades, including during its 1985 incorporation, Fourth of July parades, and the creation of the Historic Landmark Committee, Sister City Program, and Friends of the Joaquin Moraga Adobe. (*Orinda News.*)

PHOTOGRAPHER EXTRAORDINAIRE. During the 1940s and 1950s, Les Sipes flew the Contra Costa traffic route for the *Oakland Tribune* every day, but on the ground, he also produced thousands of unique and quirky images. All the photographs credited to the Contra Costa County Historical Society in this book have come from its superb Les Sipes Collection and guarantee that Sipes's talent will live on, even though he has passed away. (CCCHS.)

LOUIS STEIN (1902–1996). Society owes a huge debt of gratitude to this Berkeley pharmacist's passion for local history. After a lifetime of devotedly preserving and documenting the past, Stein donated 50,000 manuscripts, documents, photographs, and newspapers—dating from 1858 to 1920—to the Contra Costa County Historical Society, while many other organizations received gifts according to their specialties. Stein (right) is pictured with Charles Allen at Wagner Ranch in 1950.

INDEX

Adobe, Joaquin Moraga, 14, 15, 17, 122
Avenida de Orinda, 87, 89, 107
BART, 8, 44, 52, 67–70, 124
Bicentennial mural, 122, 123
Bickerstaff, Jennie, 72
Black's Market, 108
Brobeck family, 120
Brookbank Ranch, 28–31
Bryant's Corner station, 52, 54, 56, 62
Caldecott (Broadway) Tunnel, 18, 44, 49, 59–61, 63–66
California & Nevada Railroad, frontispiece, 50, 52–56, 102, 105
Camron, Alice and William, 20, 112
Casa Azul, 99, 118, 119
Casa Orinda, 44, 45, 103
Casa Verana, 39, 104
Charles Hill, 55
Contra Costa County, 12, 17, 39, 49, 69, 70, 123
Crossroads, 8, 44, 45, 52, 58–60, 62, 84, 102, 103, 109
De Laveaga family, 8, 34–37, 39, 41–43, 47–49, 52, 54, 56, 62, 81, 99, 100, 110
Del Rey Elementary School, 77
El Toyonal, 18, 37, 92
fire department, frontispiece, 85–89
Fish Ranch, 57, 58, 72
Glorietta Elementary School, 76
Greyhound Bus Lines, 69, 70
Haciendas del Orinda, 43, 47, 49, 100
Highway 24, 8, 52, 57, 58, 60, 62, 82
Inland Valley schools, 77
Jaeger, Toris, 114
Kennedy Tunnel, 38, 40, 41, 63, 64, 104, 117
Kirby, Peggy, 64, 79
Lake Cascade, 47–49, 81, 99
Landers, Bobbie, 122
landslides, 82, 84
Matchless Orinda, 21
Miner family, 28, 30, 31, 63
Miner Road, 34, 35, 54, 56, 84, 92, 99, 118

Miss Graham's Riding Academy, 101, 110
Moraga, Lt. José Joaquin, 12, 14
Moraga School, 72
Nelson family, 116, 117
Old Yellow House, 72, 116, 117
Orinda Community Center, 40, 74, 111, 124
Orinda Country Club, 11, 41, 56, 81, 99, 100
Orinda Historical Society, 10, 54, 124
Orinda Park, 20, 21, 23, 50, 52, 53, 112
Orinda Park School, 18, 26, 73, 105
Orinda Theatre, 59, 94–97
Pellissier fallout shelter, 91
Phair's, 41, 42, 107
Pony Express, 32, 62
rodeo, 98
Saklan Indians, 10
Scouts, Boys and Girls, 54, 111, 112
Sleepy Hollow, 28, 31, 80, 84, 92, 120
Sleepy Hollow Elementary School, 76
telephone directory, 121
Wagner Ranch, School, 26, 73, 75, 123
Wagner, Theodore, 22–26, 73, 105, 114
White III, William, 16, 78
Wright, Frank Lloyd, 115

About the Orinda Historical Society

Established in 1970, the Orinda Historical Society is an all-volunteer, membership-based nonprofit organization that seeks to serve the community by preserving Orinda's rich past through historical artifacts, documents, and images. The society publishes a members' newsletter and offers educational walking tours and lectures on related subjects. It is located on the plaza level of the Orinda Library building. Email info@history.org or visit the society's website at orindahistory.org to learn more.

DISCOVER THOUSANDS OF LOCAL HISTORY BOOKS FEATURING MILLIONS OF VINTAGE IMAGES

Arcadia Publishing, the leading local history publisher in the United States, is committed to making history accessible and meaningful through publishing books that celebrate and preserve the heritage of America's people and places.

Find more books like this at
www.arcadiapublishing.com

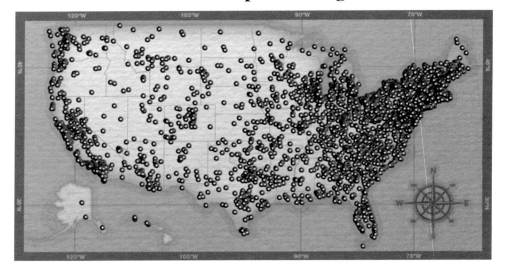

Search for your hometown history, your old
stomping grounds, and even your favorite sports team.

Consistent with our mission to preserve history on a local level, this book was printed in South Carolina on American-made paper and manufactured entirely in the United States. Products carrying the accredited Forest Stewardship Council (FSC) label are printed on 100 percent FSC-certified paper.

MADE IN THE
USA